"Climate change is reorganising work from the scale of 'the economy' to 'the body' and everything in between. Yet what these intersecting changes mean in practice for workers and communities requ[illegible] of the array of occupations, places and visions a[illegible]e the work landscape. In *Working Throug*[illegible]d Stein assemble 11 beautifully written cha[illegible]n that illuminates the informal as well as for[illegible]n-agining involved in work-based climate ac[illegible]k under the hood' that is needed to accelerate the just transition and climate change adaptation."

Professor Lauren Rickards, *Director, La Trobe Climate Change Adaptation Lab*

WORKING THROUGH PLANETARY BREAKDOWN

This book offers a genuinely interdisciplinary engagement with the future of paid and unpaid work in the context of the twin challenges of decarbonisation and the growing impacts of an unstable climate.

It is innovative in its grounding of such discussions in the everyday realities of workers' experiences with an empirical focus on skill, occupational shifts and technological change at the workplace level. *Part I: Skills and Training* delves into how workers gain crucial skills across their lifetimes. From survivalist "preppers" to local microgrid operators, the chapters reveal practical and often unrecognised but essential expertise. Case studies include air-conditioning technical educators and construction trades leveraging tacit knowledge of sustainable practices. *Part II: Industrial Transformation* draws on empirical studies from coal mining, manufacturing, defence and construction to highlight workers' experiences of climate shifts, heat and industrial transition. Theoretical contributions explore novel legal strategies such as fossil fuel "cessation" and examine the role of health and safety frameworks in addressing worker democracy and climate-change mitigation.

This collection will resonate with scholars, students, policymakers and trade unionists interested in environmental labour studies, just transitions and the future of work. It offers vital lessons for navigating complex industrial transformations.

Key features:

- Detailed case studies in critical sectors such as energy, construction, defence and manufacturing;
- A dynamic interdisciplinary fusion of human geography, political economy, sociology, industrial relations and law;
- Emphasis on worker agency, practical skill and grassroots adaptability amid intensifying climate impacts.

Chantel Carr is an Australian Research Council DECRA Fellow in Geography and Sustainability at the University of Wollongong, Australia. Chantel is a human geographer working on the social and labour dimensions of energy transitions and decarbonisation.

Jesse Adams Stein is a Senior Lecturer in the School of Design at the University of Technology Sydney (UTS), Australia. Jesse is an interdisciplinary design researcher whose work focuses on labour, technology and industries in transition.

Routledge Explorations in Environmental Studies

UN Human Rights Institutions and the Environment
Synergies, Challenges, Trajectories
Sumudu Atapattu

Globalization, Environmental Law, and Sustainable Development in the Global South
Challenges for Implementation
Kirk W. Junker and Paolo Davide Farah

Myth and Environmentalism
Arts of Resilience for a Damaged Planet
Esther Sánchez-Pardo and María Porras Sánchez

Christian Environmentalism and Human Responsibility in the 21st Century
Questions of Stewardship and Accountability
Katherine M. Quinsey

Ecological Democracy
Caring for the Earth in the Anthropocene
Odin Lysaker

Working Through Planetary Breakdown
Labour, Skill and the Changing Climate
Chantel Carr and Jesse Adams Stein

WORKING THROUGH PLANETARY BREAKDOWN

Labour, Skill and the Changing Climate

Edited by Chantel Carr and Jesse Adams Stein

Routledge
Taylor & Francis Group
LONDON AND NEW YORK

Cover image by Lizzie Buckmaster Dove, Bushfire damaged "You are Here" National Parks signage, Wog Wog campground, Morton National Park, 2020, following the Black Summer Bushfires 2019-2020, Australia. Reproduced with permission. See this book's frontmatter for artist's statement.

First published 2026
by Routledge
4 Park Square, Milton Park, Abingdon, Oxon OX14 4RN

and by Routledge
605 Third Avenue, New York, NY 10158

Routledge is an imprint of the Taylor & Francis Group, an informa business

Open Access is funded by Jesse Adams Stein's Australian Research Council Discovery Early Career Research Award (DE210100158)

British Library Cataloguing-in-Publication Data
A catalogue record for this book is available from the British Library

ISBN: 978-1-032-84325-4 (hbk)
ISBN: 978-1-032-81771-2 (pbk)
ISBN: 978-1-003-51224-0 (ebk)

DOI: 10.4324/9781003512240

Typeset in Times New Roman
by SPi Technologies India Pvt Ltd (Straive)

CONTENTS

A NOTE ON THIS BOOK'S ORIGINS

All Hands on Deck 2023

This edited collection emerged out of the cross-disciplinary symposium, *All Hands on Deck*, held at the University of Technology Sydney, 19–21 July 2023. Framed around questions of work, skill and material production at a time of human and environmental calamity, the presented papers inspired us to bring together not just one, but two edited collections that emerged from this event.

All Hands on Deck grew from our shared concern that the problems of climate mitigation, decarbonisation and industry transitions required more insights at the level of everyday human experience. We perceived a need for discourse that is more grounded, more pragmatic, geographically specific and politically nuanced, and for work that relies far less on notions of "innovative" technological solutionism or market mechanisms to envision change.

The symposium was an event like none other: the warmth, collegiality and cross-disciplinary listening that took place was truly remarkable and it was certainly "made" by the generosity and thoughtfulness of all who attended over those three days. It brought together around 150 participants in person and online, including geographers, sociologists, legal scholars, labour historians, political economists, anthropologists, design scholars, creative industries researchers, union researchers and industry policy researchers (among others). In particular, the nuances of Australian labour transitions resonated with audiences and delegates, as did questions of the ongoing need for practical skill, technical knowledge and high-quality vocational training.

Building on the energy and ideas developed over the three days of *All Hands on Deck* in 2023, this collection of essays begins the work of drawing together labour-transition discussions across a number of disciplines. This book joins its sibling collection: *Designing through Planetary Breakdown: Locating Material Knowledge and Practical Skill* (Routledge 2025).

The symposium's unusual position at the intersection of design, labour studies and the environment attracted a diverse range of perspectives and methodological approaches for examining the labour of climate change. The chapters in this book are not united by theoretical orthodoxy, but by their commitment to empirical research, lived experience and place-based understanding. We see this heterogeneity as a strength. Rather than imposing a predetermined conceptual or disciplinary framing, these themes emerged organically through the symposium's focus on everyday experience and worker perspectives. The result is a collection that – we hope – captures the diversity, complexity and materiality of working lives in a time of climate change.

We acknowledge that many of these academic conversations are ongoing, and they sustain and nourish us, in spite of the persistent onslaught of destructive neoliberal managerialism in tertiary education.

Chantel Carr and Jesse Adams Stein

CONTRIBUTORS

Editor biographies

Chantel Carr is a human geographer and planner, specialising in the social and workforce dimensions of Australia's energy transition. She is a researcher in Geography and Sustainability at the University of Wollongong. Carr's work often highlights the deeply grounded expertise and place-based knowledges that shape localised responses to planning, climate and energy policy. She has led projects funded by the Australian Research Council and contract research for the Australian government, including an ARC DECRA fellowship looking at the experiences and prospects of metallurgical coal workers, and an ARC Discovery project on Port Kembla's low-carbon transformation – work that spans both sectoral transitions and place-based community responses. Carr's research has been published in leading international journals, including Progress in Human Geography, Dialogues in Human Geography, and Transactions of the Institute of British Geographers.

Jesse Adams Stein is an interdisciplinary design researcher specialising in the relationship between technology, work and material culture. She is a Senior Lecturer in the School of Design at the University of Technology Sydney. Her research shifts between historical and contemporary contexts and focuses on the less visible sides of design: industrial craft, manufacturing, repair, skill loss and labour experiences in the face of economic restructuring and technological upheaval. She recently completed an ARC DECRA Fellowship examining the connectivity between local manufacturing production, design education and vocational training in Australia, focusing on the 1980s to the present. Stein is the author of *Industrial Craft in Australia* (2021) and *Hot Metal* (2016). With Chantel Carr, Stein founded and directed the interdisciplinary symposium *All Hands on Deck* 2023, which was the basis for this edited collection and its sister collection, *Designing Through Planetary Breakdown* (also Routledge).

Contributor biographies

Kezia Barker is a Senior Lecturer in Geography at Liverpool John Moores University. Her research considers the intersection between environmental futures and cultures of nature, particularly public responses to environmental insecurity and framings of biological threat. Kezia is interested in how wider discourses of environmental crisis are embedded and reworked in identities and everyday practices. This has included research contributions on anticipatory geographies and preparedness subjectivities. She has recently been researching prepping, bushcraft and foraging in the UK.

Tom Barnes is an Associate Professor and Principal Research Fellow, Institute for Humanities and Social Sciences at Australian Catholic University. He is a sociologist of work and economy. His research focuses primarily on insecure work, labour markets and new technologies. He is currently researching logistics and automotive manufacturing. He has written two books, *Informal Labour in Urban India* (Routledge, 2015) and *Making Cars in the New India* and is currently completing his third book on just transitions and the demise of Australia's auto manufacturing industry. His articles have appeared in many journals, including *The Sociological Review*, *Environment & Planning A*, and *Urban Studies*.

Rosie Cox is a Professor of Geography and Executive Dean of the Faculty of Social Sciences at the University of Sussex. Her research revolves around the relationship between home, work and consumption. It interrogates home as a space of paid and unpaid work and examines how work and consumption at home relate to broader inequalities of class, gender and race/ ethnicity. Recently she has been researching male labour in the home and particularly the work of home renovation, whether it is done as "DIY" by homeowners themselves or by paid "handymen." *Home Improvement in Aotearoa New Zealand and the UK* was published in 2021, by Routledge.

Daniel Daly is an Honorary Senior Research Fellow with the Sustainable Buildings Research Centre (SBRC) at the University of Wollongong. His work focuses on improving the performance and sustainability of existing buildings. Daniel has expertise in the energy efficiency of residential and commercial buildings and has authored guides to retrofitting owner-occupied and social housing.

Matthew Daly is a Research Principal with the Institute for Sustainable Futures (ISF) at the University of Technology Sydney. His work focuses on interdisciplinary research tackling complex and systemic problems in the housing, energy and sustainable built environment space.

Mark Dean is an Adjunct Researcher at RMIT University's Business and Human Rights Centre. Mark previously worked as a research and policy officer at the Australian Manufacturing Workers' Union. Mark has also held research roles at the Carmichael Centre, within the Centre for Future Work and at Flinders University, researching and teaching on topics including the future of work, industrialisation, just transition and industrial policy.

Frances Flanagan is a Lecturer in the Faculty of Law, University of Technology Sydney. Her research focuses the intersections between labour law, environmental crisis and changing formations of technology, gender and migration. She has an interdisciplinary background in law and history and has published in academic journals including the *Sydney Law Review*, *New Technology Work and Employment*, and the *Journal of International Labour Law* and *Industrial Relations*. Prior to commencing at UTS in 2023, Flanagan held academic roles at the University of Sydney and Birkbeck University of London. Outside academia, Frances served as National Director of Research at United Voice (now the United Workers Union) and was a lawyer at the Yamatji Marlpa Aboriginal Corporation, the native title representative body for the Pilbara, Murchison and Gascoyne regions of Western Australia.

Brett Heino is a Senior Lecturer in the Faculty of Law at the University of Technology of Sydney. His research interests include literary geography, the political economy of law (with a focus on labour law) and the legal and spatial structure of post-World War II Australian capitalism. He is the author of *Space, Place and Capitalism: The Literary Geographies of The Unknown Industrial Prisoner* (2021) and *Regulation Theory and Australian Capitalism: Rethinking Social Justice and Labour Law* (2017).

Elizabeth Humphrys is a political economist specialising in labour and work, and the Head of Discipline of Social and Political Sciences at the University of Technology Sydney. She researches the impact of economic and climate change on workers, and control within the labour process. She is an Editor of *Social Movement Studies*, and an Associate of the Centre for Future Work at The Australia Institute. Her first book, *How Labour Built Neoliberalism*, was described in the Sydney Review of Books as a "tremendously important" contribution to understanding economic change in Australia's recent past.

Natasha Larkin is a climate and energy policy researcher at the Centre for Sustainable Development Reform, University of NSW. With a professional background in law and a recently completed PhD in legal geography focusing on the emergence of offshore wind development in Australia, she brings an interdisciplinary legal-social sciences perspective to addressing the complex regulatory and spatial challenges of energy transition.

Kathryn Lucas-Healey is a Senior Climate and Energy Adviser at Environment Victoria, an environmental non-government organisation based in Melbourne, Australia. The work presented in this book stems from energy social science research carried out at the Battery Storage and Grid Integration Program at the Australian National University. Kat has worked in energy and sustainability for 20 years, spanning practice, policy and research. Her research examines the intersections of new and/or passive energy technologies, people and institutions.

Ben Lyall is a Lecturer in Digital Communication at RMIT University and an affiliate of Monash University (both in Naarm/Melbourne, Australia). He is a digital sociologist interested in how social life is impacted by digital infrastructure and smart devices. His work cuts across technology, wellbeing, employment, and civic life. Projects include AUTOWORK, which explores digitised, automated, and robotised work futures in retail, health, and construction sectors.

Pauline McGuirk is Professor of Urban Geography (Honorary) at the University of Wollongong, NSW. She was Director of UOW's Australian Centre for Culture, Environment, Society and Space from 2017 to 2025. As an urban political geographer her research has explored shifting modes and practices of urban governance, including a focus on carbon governance, energy transition and smart/AI cities. Pauline is a Fellow of the Academy of Social Sciences Australia (ASSA).

Daniel Nicholson is a Lecturer in Employment Relations and HRM at the University of Western Australia. His research is focused on how workers can act collectively to shape the future of their work in the context of rapid technological change and the ecological crisis. He holds a PhD from the Employment Research Unit at Cardiff Business School and worked as a Policy and Research Officer at the Australian Manufacturing Workers' Union between 2021 and 2024.

Sarah Pink is an award-winning design and futures anthropologist and documentary filmmaker, an ARC Laureate Fellow, and Professor and founding Director of the Emerging Technologies Research Lab and FUTURES Hub at Monash University. She is known globally for her interdisciplinary research and engagement. Her work brings design and futures anthropology into collaboration with other disciplines and partners inside and outside academia to ensure that everyday human and environmental futures are ethically accounted for in the design, application and use of emerging technologies. Sarah's recent publications include the book *Emerging Technologies: Life at the Edge of the Future* (2023) and her documentary *Air Futures* (2024).

Hedda Ransan-Cooper is a former academic, now researcher in the not-for-profit sector. She's interested in collective and individual re-imagining around what it means to care for each other and for place. She's inspired by scholarship, poetry and art that speak to themes of collaborative survival from the margins; rest as resistance and joy as essential to connection and transformation.

Michael Rawling is an Associate Professor at the Faculty of Law, University of Technology Sydney. He has more than 15 years' experience researching and teaching labour law. A key focus of his research has been how to regulate business models (such as supply chains and gig economy arrangements) based on precarious work and deployed to evade employment law. He has published widely on his main research interests which include the scope of labour law, labour law enforcement issues, and social democratic approaches to labour law.

Eugene Schofield-Georgeson is a Senior Lecturer at the Faculty of Law, University of Technology Sydney. He researches labour law and possibilities for transition from neoliberalism to social democratic alternatives and institutions. His most recent book is *Contract, Labour Law and the Realities of Working Life* (Routledge).

Elyse Stanes is a human geographer and ARC DECRA Fellow at the University of Wollongong. Her research examines how materials move unpredictably through social and technical systems, resisting efforts to manage them. She collaborates with industry, government, and communities to understand these dynamics, particularly in waste sectors and built environments.

ACKNOWLEDGEMENTS

The editors would first like to acknowledge the authors who contributed to this book and to its sister collection, *Designing through Planetary Breakdown*. It has been a privilege to work with such an engaged and insightful group of scholars, whose rigour and diverse perspectives have enriched the editorial process immeasurably.

Second, we would like to acknowledge everyone involved in the *All Hands on Deck* symposium in July 2023 in Sydney. We also have some more specific acknowledgments to those who particularly supported us, and worked with us, providing advice, assistance, and support along the way. Thank you to Barbara Adams, Enya Moore, Anna Cristina Pertierra, Lizzie Buckmaster Dove, Jennifer Macey, Cameron Tonkinwise, Damian White, Mary Spongberg, Jeanti St Clair, Adele Wessell, Ana Petidis, Kate Sweetapple, Zoë Sadokierski, Michelle Montgomery, Monica Monin, Jack Sawyer, Michael Fisher, Thea Kable, Sam Wallman, and the *Makers, Manufacturers and Designers* interviewees.

Thank you to Paul Allatson for his diligent editorial work, and to the Routledge team: Annabelle Harris, Rosie Anderson, Martha Uberg, and Jyotsna Gurung.

As always, we extend our personal thanks to Kevin Brand and Eugene Schofield-Georgeson, and Elliot, Louis, Archie and Renzo. The enduring memory of the late, great Paul Leon Stein AM has helped imbue these pages with care, rigour and environmental commitment.

All Hands on Deck and its publication outcomes were made possible by Jesse Adams Stein's Australian Research Council (ARC) Discovery Early Career Research Award (DE210100158), the School of Design at the University of Technology Sydney, and the Australian Centre for Culture, Environment, Society & Space (ACCESS) at the University of Wollongong.

COVER PHOTOGRAPH: ARTIST'S STATEMENT

Lizzie Buckmaster Dove, *You Are Here*, 2020

The cover photograph is an image of bushfire damaged "You are Here" National Parks signage, Wog Wog campground, Morton National Park, 2020, following the Black Summer bushfires 2019–2020, Australia.

These photographs are from a series I took of National Parks signage. These signs typically tell the visitor about the place where they are. Prior to being melted by fire, they told of flora and fauna species, gave some details about topography and geology, and included a simple map with other details about geographical features. These signs tell the viewer where they are. Often on this kind of information board, there is an arrow in the map pointing to exactly where the viewer stands. It is often red or in bold, and reads, "You are here."

I encountered the first sign in 2020 when wandering through Conjola National Park, north of North Bendalong, New South Wales. The sign with the melted part extending below the frame was at the carpark to Monument Beach. I … walked there along the beach, past all the hastily erected "do not enter the national park" corflute signage. I was barefoot and was savouring the soft sand and sandy soil beneath my feet. I first saw the sign in March (before pandemic made it impossible to be there). … When I returned, someone had picked away parts of it to reveal the kangaroo. The second sign I photographed – the one used on this cover – is from Wog Wog campground and was photographed in that same year.

When I first came upon these burnt signs, I remember having a very visceral response. I could no longer see anything of the traditional wayfaring details on the panel. Everything was obscured by the strange and other-worldly melted plastic that once protected the paper sign from weather; now its own kind of topography. It spoke to me of another kind of landscape that until then had been present to me only in intellectual, peripheral and future-imagined ways. A changed landscape resulting from one of the many possible exertions of an extreme event, resulting from climate change.

I guess I wasn't "thinking" this at the time, so much as feeling it. What I felt was very much that the signage was still and more astutely than ever perhaps, telling me where I was standing in place/time. In the newly iterated "You are Here", what had changed was that the "here" was no longer imagined, peripheral or in the future, but "here, and now."

In this new manifestation the sign was doing some truth-telling that was un-uttered by the then, climate-change denying Morrison-led government. The truth-telling-sign was speaking for the local place of Conjola National Park, and the geographical expanse of the eastern coast of Australia that had borne the Black Summer fires. The plastic protecting the paper map from weather had become weathered by an event resulting from climate change. Now it was reading and telling the true state of weather and the climate. A new post-climate-change synoptic chart. The sign was also speaking for a climatic global place: You are Here. We are indeed.

I was fascinated, repelled and compelled by the signage. It was so tactile, so interesting to touch. I stood there alone in a mostly silent, razed landscape, where my walking was unimpeded by undergrowth. I could walk anywhere. Charred ground snaked through my toes. I could have been the last human on earth.

Lizzie Buckmaster Dove, 2025

SKILL, INDUSTRIAL TRANSFORMATION AND WORK IN A CLIMATE-CHANGING WORLD

Chantel Carr and Jesse Adams Stein

Introduction

In early 2025, images of catastrophic unseasonal wildfires bearing down on Los Angeles, California, shocked the world. The intensity was clearly evident as the fire moved rapidly through the city's hills and reached deep into its iconic beachside suburbs. On the other side of the Pacific Ocean and in a different hemisphere, east coast Australians were watching closely. Many knew exactly what the city's residents were facing: the scenes were reminiscent of the 2019–2020 Black Summer bushfires that devastated south-eastern Australia, destroying 3000 homes and killing 33 people.[1] There was also deep concern among Australia's firefighting experts for what these Californian winter fires meant for local firefighting capacity. Australia leases up to six large water-bombing tankers from North America during the southern hemisphere summer. Former Fire and Rescue NSW Commissioner Greg Mullins outlined the demands an extended North American fire season would place on the highly skilled specialists whose services are now in demand across two continents *at the same time*:

> Fire season used to be quite distinct. The North American fire season would finish and ours would start. We'd be able to share things like large aircraft and if we had big fires, we could call them and get specialist firefighters, incident management personnel.[2]

Mullins' insight starkly illustrates how the changing climate is already disrupting established patterns of work, in this case resource sharing for emergency response between two nations. The example highlights critical questions about the capacity of existing systems to cope with the escalating challenges of a warming world.

DOI: 10.4324/9781003512240-1

It is evident by now that climate change is already impacting working lives, and that it will dramatically re-shape the future of work across all sectors.[3] At the heart of mitigation efforts is the imperative to decarbonise energy generation, industrial production systems, the built environment, transport, food and agriculture, and waste systems.[4] Enormous amounts of human labour and skill are urgently required to plan and implement these transitions. Such large-scale decarbonisation efforts are necessary for lowering greenhouse gas (GHG) emissions to a level that will – in a best-case scenario – limit the catastrophic climate impacts of a global surface temperature that has already broken through a 1.5°C threshold. However, industry transitions and decarbonisation are not the only contexts where work and skills emerge in climate change response, and arguably they are not always the most important. As the example above shows, severe weather patterns, rising sea levels and unprecedented weather disasters across the world are generating additional forms of paid and unpaid work. This is the work of climate adaptation: responding to extreme conditions, moving populations away from places in harm's way, rebuilding infrastructures (both social and physical), and supporting communities recovering from disasters.[5] Much of this work is highly practical and localised in nature: it is directly responsive to a given moment, it is often manual or physical, it can be dangerous, it is often social, organisational, highly complex, and based on care for people, ecologies, infrastructures and environments.

While climate science has given us a reasonably clear picture of the environmental realities we are likely to face, less is understood about the skills and capacities that will be needed even in the very near future, or how existing labour structures and work tasks will be transformed. Across the spectrum of climate change mitigation and adaptation, research is urgently needed to understand how the very nature of work itself is changing, and how we might work through the challenges ahead.[6] Inevitably, climate change will demand new ways of theorising and understanding work.[7] Accordingly, the way scholars, policymakers and labour institutions respond to questions about human labour must also shift and expand. How might the work of climate mitigation and adaptation be resourced and organised? What impact will technological change have on both established and emergent labour processes? How might we need to rethink skills and training to address decarbonisation and disasters simultaneously? Who will benefit from the wholesale re-organisation of work that is already underway, and who might get left behind? How can we ensure that work in a warming world does not reproduce the injustices – colonialism, rampant resource extraction, and exploitation of vulnerable social groups – that have characterised capitalist economic development to date? What role can workers play in shaping climate change mitigation and adaptation? And is it even fair to ask that of workers – given the chief culpability for determinedly increasing GHG emissions in recent decades lies with a relatively small number of multinational corporations?

Such questions – endless in themselves – feel very 'of the moment', but there is evidently an extensive legacy of research on work, labour and skill that has examined similar themes in earlier moments of industrial transformation and upheaval.[8] While the authors in this collection write with their gaze on current and future challenges, we remain deeply informed by labour theories, debates and research from past decades, across the social sciences, law, history and industrial relations. Emerging out of nineteenth and twentieth century industrialisation, questions of worker agency, power relations and collective action have energised progressive labour researchers and activists for decades. Much of this work has been pioneered by feminist scholars, whose research has radically expanded perspectives on what counts as work, and contributed new insights into how different kinds of work are valued through intersecting lenses of gender, class and race.[9] Such approaches have long drawn attention to how capitalist economies rest precariously on a bedrock of unpaid, undervalued and largely invisible work, much of it centred on care.

Climate change is evidently scaling up the need for care. Patrick Reid, Chief Executive Officer of Australian aged care provider IRT Group vividly described the multiple challenges their workers faced in keeping residents on the New South Wales (NSW) south coast safe during the Black Summer bushfires:

> Transferring frail aged care centre residents was a significant challenge – a number of residents required ambulance transfers at a time when the ambulance service was under significant pressure. [The] employees have been simply amazing. Despite the loss of mains power, road closures and communication blackouts, they have focussed on caring for our residents and keeping them calm, sometimes while their own homes and families were under threat.[10]

Reid's account illustrates not only how care is crucial in disaster response, but also the complex intersections of paid and unpaid labour, and how geographical context and localised knowledges are central to the work of climate adaptation. In reflecting on such accounts, long legacies of labour research remind us that it is critically important that we remain attentive to worker agency and intersectional power dynamics. As we seek to understand how working lives are re-shaped by new waves of change, we must build carefully on existing research foundations, remaining ever-mindful that work extends beyond paid employment, varies across diverse geographies, and that it is *always* politicised.

The *All Hands on Deck* project[11] – our 2023 symposium and two edited collections – has sought to bring a sense of historical and geographical sensitivity to bear on the pressing work of climate change, and the skills that will be required to address it.[12] This collection – *Working through Planetary Breakdown* – is purposely diverse in approach. While some authors are established labour and industrial relations scholars, others have approached questions of work and skill through their research on climate and energy policy, design, law, and political economy. Some contributors – such as Mark Dean and Daniel Nicholson (Chapter 10), Elizabeth Humphrys (Chapter 9) and Tom Barnes (Chapter 11) – are explicit that climate change is not the only challenge confronting the domain of work, exploring intersections with other current global 'breakdowns', such as the fallout from the Covid-19 pandemic and the opportunities and challenges of recent technological transformations ('industry 4.0').

This introduction sets out a research agenda that is centred on a grounded approach to questions of work, skill and climate change. It makes the case that important insights on the creative and adaptive capacities of workers emerge when starting from the perspective of lived experience. These capacities will be crucial for meeting the challenges ahead, but they are almost impossible to locate and support without detailed qualitative research. While labour and economic statistics are instructive and can tell us a great deal about broader geographical or sectoral shifts, scaling down to the 'fine-grain' is critical for understanding how people anticipate and manage all kinds of change, both in their working lives and of course, outside of it. Accordingly, we have placed an emphasis in this volume on empirical accounts of work and the perspectives of workers, with eight of the eleven chapters engaging with themes of skill adaptation, occupational shifts, and lived experience amidst environmental and technological change. The remaining three chapters are broader in scope, but no less important: they offer practical critique and concrete pathways on some of the most pressing concerns at the intersection of work and climate today: just transitions and the imperative to cease fossil fuel extraction, under-utilised routes for revitalising industrial democracy and worker voice, and the need to integrate industrial policy with workforce development at the regional scale.

In what remains of this introduction, we situate our grounded approach in relation to existing and emerging research on climate change, work and labour. We emphasise the importance of deeply engaging with people and skills both within and beyond existing labour markets, and connecting these insights with the starkly uneven ways climate change and the dynamics of transition are unfolding across geographies. Our launch point is the now well-established just transitions discourse, which is central to the uneven landscape of climate mitigation response – from the necessary transformation of carbon-intensive sectors to the scaling up of clean energy industries.[13] We then turn our focus to the even more urgent work of climate change adaptation, which has received far less attention in Global North settings.[14] Emergent in this context are questions of what and whose knowledge is brought to bear to manage the uneven – and already present – risks and vulnerabilities in the context of climate change. Our interests here are in how we might identify and re-value marginalised, vernacular and place-based knowledges and skills in addressing climate adaptation, and how the agency and resourcefulness of workers *already on the front line* might be better integrated into adaptation governance. We then reflect on what it means to do research on work and climate change from Australia, a wealthy country with a heavily resource-based economy, where the effects of climate change are also already evident in catastrophic fires and floods in recent years. Finally, we outline the structure of this collection, and provide an overview of the chapters within.

The work of climate mitigation: Just transitions, decarbonisation and industrial transformation

How do we ensure that no one is left behind through the structural transformation that will accompany decarbonisation? This question is at the heart of *just transitions*, a contemporary labour project encompassing key aspects of the debate around work and climate change mitigation. With its roots in North American unions active in the chemicals sector in the early 1990s, the concept of a just transition emerged out of the comparative neglect of workers and communities affected by the regulated closure of environmentally damaging industries.[15] In contrast with attention directed towards the rehabilitation of toxic land, Tony Mazzochi of the US Oil, Chemical and Atomic Workers Union (OCAW) argued that workers and their families were neither recognised nor compensated as production ceased. In the 30 years since OCAW began advocating for a just transition for chemical industry workers, the concept has found application across many sectors. At its core, institutional proponents of a just transition (i.e. organised labour) recognise that as governments and private capital shift away from carbon-intensive industries, workers who are not as mobile – occupationally or geographically – deserve a voice, genuine agency, and in some cases material assistance to transition to new kinds of work.

With substantial union, academic and activist advocacy across different sectors, the political purchase of just transitions has greatly accelerated over the last 10 years, to the point where the concept is now used broadly, often as shorthand for seeking ethical working conditions and more democratic worker representation across many sectors.[16] The International Labour Organisation's *Guidelines for a Just Transition towards Environmentally Sustainable Economies and Societies for All* was produced in 2015.[17] These guidelines recognise that workers, governments and employers are all key stakeholders in developing transition plans that consider the environmental impacts of some sectors, while also addressing the need for decent work and social inclusion. The concept of a just transition also gained further reach in 2015, when it was included in the preamble of the Paris Agreement: "Taking into account the imperatives of a just transition of the workforce and the creation of decent work and quality jobs in accordance with nationally defined development priorities." The embrace of just transitions by global institutions has had mixed consequences, not least that the concept has become increasingly slippery, and is sometimes deployed in disingenuous ways so that extractive business-as-usual can continue.[18] In this collection, Chapter 6 sees interdisciplinary labour law scholar Frances Flanagan contribute to these discussions with a proposal for a simple and direct intervention that would give the concept some "teeth". As Flanagan argues, a key weakness of just transitions frameworks are that they often lack a clear remit to cease fossil fuel production. She proposes adding a "just cessation" clause to institutional transition frameworks, explicitly calling for an end to fossil fuel extraction *with specified time frames*. This would, Flanagan argues, enable employers, the state, and labour institutions to more directly address the implications of fossil fuel phaseout for workers, taking at least some of the edge off their conflicting interests in the process, while prioritising material outcomes for both the environment and workers themselves.

Meanwhile, others continue to question the often-limited application of just transitions on supporting workers in carbon-intensive industries, arguing that everyone is going to be affected by climate change and the industrial transformation that is already underway. We remain mindful that many workers are often rendered invisible by an over-emphasis on energy systems or industrial production. Workers in smallholder agriculture and food production, for example, including many in Global South/majority world settings, are disproportionately impacted by the effects of the changing climate.[19] Feminist and critical researchers and activists also caution against an over-emphasis on industrial workers, noting the risks of reinforcing existing hegemonies – many marginalised groups are unlikely to have access to the vast numbers of green jobs anticipated by powerful organised labour institutions.[20] Much of our own research has centred on industrial workers, although we note that this is conducted with a gender-reflexive lens that seeks, for example, to understand the operations of hegemonic masculinity in workplaces, and the motivations behind particular political persuasions.[21] Accordingly, we share many of these sentiments regarding invisible labour, and can certainly testify to the gendered, classed and racialised nature of industrial work in many Global North contexts.

Beyond sectoral perspectives we foresee a critical *place-based* dynamic to transition and industrial decarbonisation that is receiving growing attention, but has not yet landed in the intimate spaces of everyday life beyond broadly generalised 'affected workers and communities'.[22] It is worth explicitly remembering that workers do not exist in a vacuum independent of other social relations. As Carr and Larkin discuss in Chapter 7, industrial workers are embedded in households, families and communities where other household members are frequently employed in service sectors.[23] *Their* lives too, are already being reshaped as key industrial sectors decarbonise. In short, while we see continued traction in the idea of just transitions, we join many colleagues in advocating for more care and specificity in how these ideas are deployed in practice across different contexts.

We are however encouraged by nuanced perspectives on just transitions emerging from the field, which are already testing the limits of conceptual and discursive approaches.[24] While global and national frameworks are important analytical entry points, 'the ground' is where explanation, expressions of solidarity and most importantly, workers and communities, reside. Here we look to advocate for more empirically grounded research that looks deeply – and *qualitatively* – at the skills and capacities people already have, as well as those needed, to address climate mitigation and adaptation. Such research is vital for moving beyond narrow technical framings of skills as mere labour market commodities that can be exchanged or transferred between sectors.[25] Rather, skills are complex, situated, sometimes invisible and often undervalued – they fundamentally give people agency, not only as paid workers within labour markets, but also beyond them.[26] It is precisely in looking beyond the technical competencies often invoked in institutional just transitions frameworks that adaptive capacities and crucial place-based knowledges emerge. The first section of this volume, *Skills and Training*, looks at these themes in detail, addressing the diverse ways people acquire skills, the role of informal learning and embodied knowledge, and how skills are transferred and deployed across domains outside of paid work, including household adaptation, community-based skill sharing, and forms of practical knowledge that often go unrecognised in formal transition planning. Such skills and informal networks, while rarely captured in policy discussions, are already proving crucial for community resilience and adaptation.

Across the wealthy economies of the Global North, the last two decades of climate policymaking has emphasised sectoral decarbonisation and the challenge of climate change mitigation, yet this attention has not translated into sufficient action.[27] It is now clear that adapting to a changed climate is both necessary and urgent, and that communities worldwide are already experiencing vastly uneven capacities to respond and cope with these changes.[28] Inadequate planning is starkly evident across both mitigation and adaptation domains: in the hasty withdrawal of global capital that leaves local workers, communities and governments scrambling to respond; and in the volunteer labour required to plug the gaps in the aftermath of disaster events. This largely reactive approach has significant risk implications for workers and communities. In an analysis of adaptation-related job advertisements in Australia, Todd Denham and colleagues point to both urban and professional services biases in how planning is currently conceived.[29] Their findings highlight uneven capacities for adaptation planning, particularly in non-metropolitan regions that are often also host to carbon-intensive industries on the front-line of mitigation efforts. The situation is one of a 'perfect storm' for such regions, compounding skills gaps and deepening vulnerabilities. An integrated, place-based approach to just transitions *that includes adaptation* would help address the gaps inherent in sectoral-based planning,[30] ensuring that workers and communities are not left to navigate the complex challenges of climate change with inadequate support, further deepening existing inequities.

From sectors to places: Flexible capacities, care and the work of climate adaptation

Many of the chapters in this volume emphasise workers' capacities for flexibility, in turn (and often inadvertently) highlighting the limits of existing institutional skills frameworks for addressing the challenges of climate change. In their fetishisation of job numbers and the formalisation of 'green' skills, governments at all levels frequently overlook how people move through careers and between sectors, stringing diverse skills together in ways that are difficult to capture in quantitative terms. In Chapter 5, Jesse Adams Stein provides a counterpoint to high-level and often abstract perspectives that fail to capture the portability of skills across domains. Stein's approach – examining workers' lived experiences and career trajectories – reveals capacities for transferring skills across domains that jobs metrics typically miss. The formal labour market consistently undervalues these flexible capacities often because they are simply invisible, and despite their effectiveness in 'getting the job done', for example in localised responses to disaster situations.

This disconnect between institutional conceptions of 'climate work' and the flexible capacities actually needed and deployed during disaster and disruption has significant implications for both policy and practice. When adaptation planning narrowly focuses on technical risk assessments performed by credentialed professionals (who often 'drop into' regional locations from major metropolitan centres),[31] it reproduces what Pearse and Bryant identify as "hierarchies of value"[32] that discount the essential on-the-ground labour of repair, care, and community-building. These hierarchies privilege certain forms of expertise while rendering invisible the practical skills that communities depend on during crises. As communities navigate both incremental adaptation and acute disaster response, they will come to rely heavily on people who can integrate new responsibilities into their existing roles, cross professional boundaries, and apply their experience to new and unprecedented situations. In short, the work of climate adaptation is already taking place on the ground, though it is largely overlooked in national adaptation policy and planning.

Questions of how to locate and support care are also central to understanding the work of climate adaptation, and are reflected in this volume. The systematic undervaluation of care across many societal contexts is evidently being reproduced in relation to climate change. In Chapter 9, Elizabeth Humphrys examines how extreme heat compounds existing workplace vulnerabilities in the home-based care sector. Humphrys' findings demonstrate how the burden of managing climate risks falls disproportionately on already-marginalised workers, often in low-paid roles. This care work is all but invisible, yet it will play an increasingly critical role in community resilience. Despite this systematic neglect, existing institutional frameworks offer some hope in the form of under-utilised pathways to better protect and value care in climate adaptation.

Stronger powers for Workplace Health and Safety (WHS) committees for example, as Eugene Schofield-Georgeson and colleagues identify in Chapter 8, could provide a route through which such concerns might find traction within existing legal structures. However, as the authors collectively caution, a disconnect between these protection mechanisms and the outcomes they currently deliver reveal how power imbalances shape adaptation in practice.

Critical gaps in how the work of adaptation and mitigation are conceptualised and valued underscore the essential role of grounded qualitative research that centres the experiences of workers themselves. Context matters profoundly in climate response: regional geographies shape sectoral transitions to address mitigation, while adaptation strategies must be distinctly place-based to be effective. Workers' lived experiences also reveal how communities self-organise across different scales, from workplace committees to neighbourhood networks to larger regional and national alliances, creating flexible capacities that formal planning often overlooks. In what follows, we examine why Australia is a particularly illuminating context from which to understand these dynamics.

Writing from Australia

Aside from the first chapter by Barker and Cox on UK prepper cultures, the chapters in this collection largely emerge from an Australian context. As editors we have grappled with this geographical emphasis,[33] in time coming to view it as a strength, rather than a limitation. Australia has spent a good part of the twenty-first century in the climate 'doldrums', and is now scrambling to catch up industrially, following what was a decade of climate inaction. From 2013 to 2022, as efforts to address climate change began to dominate the geopolitical agenda globally, the conservative Liberal-National Coalition held power in Australia. This was a government that actively obfuscated action on climate change. Party rhetoric around the importance of fossil fuels to jobs, and to the Australian economy, dominated political and media discussion. The third conservative Prime Minister of this term, Scott Morrison, famously brandished a lump of coal in the Australian parliament in 2017, smirking as he said to fellow parliamentarians: "This is coal. Don't be afraid, don't be scared."

After a decade in the global 'climate wilderness', Australia experienced a (slight) shift towards a more progressive climate politics in 2022, with the election of the Australian Labor Party. In recent years the nation has seen some concerted action in addressing energy transitions and the work of climate mitigation and adaptation. At the time of writing, the Labor government has just been re-elected. Their first-term track record included the *Climate Change Act 2022* (Cth) with specified emissions reductions targets, and the *Future Made in Australia Act 2024* (Cth) aimed at identifying and supporting priority sectors for investment with an emphasis on emergent 'green' industries. During this term a Net Zero Economy Authority was established, mandated through the *Net Zero Economy Authority Act 2024* (Cth), with a coordinating remit to manage transitions from coal-fired energy generation. The *Offshore Electricity Infrastructure Act 2021* (Cth) was also passed, setting out the parameters for a whole new sector to deliver gigawatt-scale renewable energy in Australian waters. But the progress has been offset by some significant missteps: the planned repeal of the *Environment Protection and Biodiversity Conservation Act 1999* (Cth) and its replacement with stronger environmental protection measures was thwarted in 2024,[34] and there remains persistent bipartisan disinterest in addressing Australia's Scope 3 emissions (largely coal and gas exports used in other nations). The challenges of addressing climate and energy while bringing the electorate along remain immense, and seemingly beyond the capacity of the two major parties in Australia. Needless to say, the space is highly contested and the political will for meaningful but difficult structural transition is patchy at best, even amongst the supposedly 'progressive' Labor party who continue to approve new coal and gas developments. Almost immediately following their re-election, Labor signed off on the North-West Shelf project, extending Woodside's licence for gas extraction until 2070. Action on climate change is made infinitely more difficult by a Liberal-National Coalition opposition intent on wedging the current Government on climate and energy. Whether the last three years of moderate progress will continue remains to be seen, and global geopolitical factors – not least the uncertainty precipitated by the re-election of Donald Trump in the US in 2025 – will also have a significant bearing on Australian climate politics over the next decade.

The authors in this collection are tracking these national and global developments closely, and with many sectors facing significant challenges around decarbonisation, there is a great deal more to be analysed. Yet in so many ways, things have not really shifted at all. Without addressing its Scope 3 emissions as one of the world's largest coal and gas exporters, Australia essentially remains a fossil fuel-reliant nation, with an economy still dominated by resource extraction. In this context, the nexus between work and the environment is somewhat reminiscent of earlier debates in North America and Europe, but it also simultaneously has its own unique flavour: one of sheer, urgent desperation set against a prevailing inertia. This makes Australia a useful litmus test for how significant industrial change might be achieved *quickly*, given the right policy settings, resources and political will (in a best-case scenario). On the other hand, the case of Australia might also be instructive to other geographies in terms of *what not to do*, should the strength of fossil-fuel industry lobbying continue to undermine meaningful reform or the just transitions agenda. Either way, it is clear that tensions reverberate across sectors such as energy, mining, manufacturing, construction and defence, which make up a substantial proportion of the country's economy. These tensions invariably carry over into the lives of workers.

The other reason Australia is an important context for labour-climate research is because of the intensity of climate disasters *already* being faced in this country. Some of the harshest impacts from the changing climate will be felt in the Pacific region. This includes wildfires of increasing frequency and ferocity, widespread flooding, extreme drought, unprecedented storms and rapid species extinction, alongside the global impacts of sea level rise. Two months prior to the 2025 Australian election, Cyclone Alfred bore down on the coast either side of the border between southern Queensland and NSW. Not only was it the first cyclone this stretch of coast had experienced in over 50 years, but residents were already weary and on edge – many were still working on rebuilding following devastating flooding in 2022 that resulted in almost 100,000 insurance claims totalling $1.33 billion in damage.[35] Two weeks after the election, the mid-NSW coast was hit by a 1-in-500-year flooding event, killing five people. Clearly, climate change is already a part of life and work for so-called 'ordinary' Australians, who experience almost daily discussions about heat mitigation, for instance, or contingency planning for fires, smoke inhalation or anticipated floods. The country is likewise characterised by a sense of urgency in terms of energy transitions and industrial transformation, but also a sense of foreboding in some regions and sectors of the community. Community conflict and disagreement about the nature of that change, fears for job losses, and concerns over skill-shortfalls, are also common elements in contemporary discourse.

The core themes of this volume – skill and industrial transformation – contain questions and issues that many regions and countries are grappling with. In particular, parallels between the Australian context and other capitalist Global North nations with a declining industrial manufacturing base and decarbonising energy sectors feature prominently. The collection offers insights that partially overlap with North American and European-focused studies, while demonstrating the particular challenges for transitions in a resource-extraction dependent nation. The Australian experience reveals the complexities of needing to simultaneously address adaptation and mitigation, especially in communities facing both immediate climate impacts and economic restructuring. As the International Labor Organisation makes clear, issues of skill-shortages, disaster recovery, 'just transitions' and manufacturing challenges are of course globally interconnected and networked, and many of the collection's authors situate their work within the context of political economy scholarship, lending an inherently international perspective to specific case studies. Ultimately, this volume highlights the importance of understanding local contexts while engaging with broader global discourses – bringing grounded, place-based insights to bear, creating just and sustainable futures for workers and communities on the frontlines of climate change.

Having worked through our approach to work and climate change and reflected on the context of writing from Australia, we now turn to a brief overview of the structure of this volume. *Working Through Planetary Breakdown* is structured in two sections: the first, "Skills and Training", looks at how workers come to acquire skills through formal education, but also by informal means, across lifetimes and in unrecognised ways. The second section, "Industrial Transformation", looks at worker-centred experiences of change in the context of climate crisis. Together, these sections reveal how workers' adaptability and practical knowledge are crucial yet often overlooked resources in navigating planetary challenges.

On Part I: skills and training

In this first part of the collection, our aim is to foreground how skills are acquired, with a view to understanding the many routes through which people come to be skilled within and beyond the workplace. In keeping with the intent of broadening the scope of how we look at work, we examine learning pathways that extend beyond formal education and training. The chapters that follow reveal how practical skills are acquired and transferred over working lifetimes, often in unexpected places and through informal networks. This approach challenges conventional assumptions about skill development, highlighting the diverse ways workers adapt to – and sometimes shape – change in their fields.

In Chapter 1, geographers Kezia Barker and Rosie Cox examine how practical skills are acquired and shared within households and communities, using the example of DIY survival preparation amongst UK-based 'preppers'. Originally emerging as a response to imagined crisis scenarios, prepping centres on practices including stockpiling food and equipment and honing survival skills. The Covid-19 pandemic, climate-related disasters and geo-political instability have since propelled prepping from sub-culture to mainstream prominence. Drawing on Barker's ongoing ethnography of online prepper groups, the chapter examines discussions of 'apocalypse skills' – those needed to survive an unspecified but catastrophic situation. The chapter tracks how these discussions often go beyond specific objects, materials and skills to something deeper – a form of *adaptability* grounded in capacities to cope with change. For Barker and Cox, the crisis framing of prepping brings both pragmatic and conceptual potential: demonstrating how resources might be repurposed while highlighting value systems that operate outside of the take-make-waste model of capital accumulation. Though seemingly extreme, the case study provides a vivid framing through which to understand the value of overlooked 'soft' skills, particularly the flexibility that comes with a practical orientation, and a capacity for creativity and imagination.

Flexible capabilities are also an underlying theme in Chapter 2, where climate and energy policy researchers Kathryn Lucas-Healey and Hedda Ransan-Cooper map out the skills needed to deliver localised energy transitions through grid-tied microgrids. Their case study is focused on the South Coast of NSW, a region ravaged by bushfires in the Black Summer of 2019–2020. Drawing on stakeholder interviews across energy, emergency management, and community sectors, the authors position their socio-technical analysis at the nexus between aspirations for localised, grid-tied microgrids, and what is needed to realise them. Their findings highlight that while local capacity building and translational skills are critical for delivering microgrids to climate-vulnerable communities, developing these capacities depends on a fundamental reorganisation of the existing centralised energy regime. Local communities have an important role to play in microgrid system design and decision-making; however existing governance frameworks fall short in creating space for energy democratisation. To address these shortcomings, the authors argue that skill development must span multiple scales – from national energy policy through to community systems and household practices.

In Chapter 3, an interdisciplinary team of engineers and social scientists, led by Elyse Stanes, critically examine the training regime that prepares 'fridgies' – skilled contractors who maintain heating, refrigeration, ventilation and air-conditioning systems – for the significant task ahead in installing, maintaining and repairing the energy-intensive cooling systems that are fundamental to human survival and comfort. Drawing on a large sectoral study, here Stanes and colleagues focus on the insights of vocational teachers, who play a crucial role in developing a skilled workforce, amidst structural constraints at the nexus of educational governance, rapidly evolving technologies, and industry needs. Through qualitative interviews, the challenges teachers face in delivering good educational outcomes become clear - outdated and overcrowded curricula, time-consuming compliance processes, and inadequate support networks. In response, teachers have developed their own networks for sharing information and professional development, effectively 'patching a gap' in preserving and developing trade knowledge. The question remains of how sustainable this 'patching' is, for a largely invisible trade on the front line of decarbonisation and energy transitions across many sectors, including the global food chain, agriculture, medicine, and the built environment.

Chapter 4, by social scientists Ben Lyall and Sarah Pink, explores another empirical case study, this time examining forms of "quiet sustainability" in the construction industry. In a sector where change is often imagined as technologically or financially driven at scale, the authors shine a light on individual practices in manual trades. Lyall and Pink's detailed accounts of the work practices of an independent designer/builder and a vocational educator reveal values centred on 'doing more with less'. This ethos is enacted through the practical work of exploring innovative materials, developing localised reuse strategies, and adopting passive design principles that respond to local climate and context. Lyall and Pink focus on the 'everydayness' of this work, showing how skilled tradespeople amplify their 'quiet sustainability' through ongoing interactions with colleagues, suppliers, educators and apprentices. This provides an important counter-perspective on where sustainability might be located, and the capacity of small-scale practices to reshape this carbon-intensive sector.

Rounding out the first section on skills and training, in Chapter 5 interdisciplinary design scholar Jesse Adams Stein examines questions of skill and flexible capacities through a different lens, drawing on oral history research with practitioners who span creative and industrial sectors. Stein uses the terms "creative technicians" and "technical creatives" to describe workers who embody diverse skillsets, often drawing skills and knowledge across occupational boundaries as they shift between designing and making. While acknowledging this flexibility often stems from the precarity deeply embedded in neoliberalised 'creative industries', Stein argues that there are valuable lessons for other contexts. In an era of climate change, technological change, and geopolitical instability, Stein provocatively shows how conventional categories of occupation, skill and education are no longer fit for purpose. Instead, Stein invites us to consider what capacities people already have, when we take the time to listen to their stories of employment and training.

On Part II: industrial transformation

In Part II: *Industrial Transformation*, we turn our attention to the industrial and technological transitions that characterise the global imperative to decarbonise. This section contributes sectoral case studies covering a range of carbon-intensive industries, including coal mining, construction, and manufacturing. Importantly, it also advances key theoretical debates around how law – specifically industrial relations and labour law, and occupational health and safety frameworks – might be used to advance climate change response in different ways.

The first of these critical theoretical contributions is the aforementioned chapter by Flanagan, who in Chapter 6 raises important questions about the adequacy of existing just transitions frameworks for grappling with the existential threat climate change presents to workers. In proposing a more specific "just cessation" statement be added to the transition frameworks and commitments used by national and international institutions, Flanagan argues that referring *directly* to fossil fuel phasedown will create firmer boundaries and re-focus transitions frameworks to materially address climate change. Flanagan situates this claim historically, reminding the reader that post-war reconstruction provided an important precedent for the scale of change required to re-orient whole economies and societies away from fossil fuel production. The chapter is also resolutely future-oriented, pointing to institutional legal forms no longer fit for purpose. Flanagan persuasively argues that labour law actively positions workers and environment in opposition, disincentivising change by protecting work in extractive sectors, and making no provision for employee bargaining on environmental grounds.

Continuing the theme of transitions with an empirical lens, in Chapter 7 Chantel Carr and Natasha Larkin take a geographical perspective, making the case for more place sensitivity in examining coal transitions. They begin with an analysis of the political economy of coal that contextualises Australia's role as a major global exporter. Drawing distinctions between two types of coal (thermal and metallurgical), and two markets (domestic and export), Carr and Larkin explore how these distinctions are rarely made in public and political discourse, yet produce vastly divergent trajectories for Australia's coal mining regions. This broad overview provides important context for their qualitative case study with metallurgical coal miners and households connected with the southern coalfields of NSW. Here they find households who have long negotiated challenging working conditions and the precarity inherent in global commodity cycles, and who piece together information about the future of the sector and their own working futures from a wide range of sources. Carr and Larkin argue that clearer policy direction would help build trust amongst workers in metallurgical coal, enabling households to plan for their futures with more certainty.

In Chapter 8, legal scholars Eugene Schofield-Georgeson, Michael Rawling and Brett Heino offer a pragmatic perspective on addressing climate change through Australia's existing labour law structure. They identify Workplace Health and Safety (WHS) committees as an under-utilised avenue through which workers might address environmental and climate-related issues through strong legal provisions widely available across Australia and the Global North more generally. While acknowledging an historical decline in the use of WHS provisions, Schofield-Georgeson, Rawling and Heino argue that realising the full scope of climate action available to workers through WHS powers, requires stronger legislative support and broader WHS reform. The authors explore how strengthening WHS avenues for worker-led climate action must sit alongside more concerted trade union involvement in environmental issues, as part of a broader strategy for strengthening industrial democracy and ultimately addressing climate change.

In Chapter 9, political economist Elizabeth Humphrys also takes up questions of workplace health and safety in the era of climate change, drawing on qualitative empirical data from her ongoing research with the United Workers Union in Australia. Humphrys highlights the compounding effects of dealing with multiple crises, finding that experiences of high heat in diverse sectors including health, cleaning and home care were further exacerbated by wearing necessary Personal Protective Equipment (PPE) during the Covid-19 pandemic. Humphrys reveals how current WHS frameworks for managing heat stress became inadequate when placed under further pressure by the implementation of Covid-19 protocols. Driven primarily by maintaining productivity and minimising additional costs, employers failed to mitigate compounded heat impacts placing workers under strain and at risk. Despite rhetoric around shared responsibilities for health and safety, Humphrys argues that power imbalances between employers and workers render climate change adaptation a political process in the workplace, with the potential to reproduce and exacerbate existing vulnerabilities and exploitation.

Chapters 10 and 11 take an oblique but complementary approach to questions of work and climate change, sharing cautionary perspectives on current and future green industrial transitions drawing on other sectors and contexts. In Chapter 10 industrial relations researchers Mark Dean and Daniel Nicholson look at barriers to workforce development through a case study of developing new naval shipbuilding capabilities in Australia. Sovereign manufacturing emerged as a policy concern following the Covid-19 pandemic and has continued to dominate Australia's industrial policy development in response to global geopolitical tensions. Meanwhile, the government is looking to energy transitions and decarbonisation to drive the generation of thousands of new manufacturing jobs, predominantly in regional areas. Drawing on union perspectives, Dean and Nicholson identify the challenges of developing a workforce to deliver new advanced manufacturing capabilities, where the need for upskilling workforces conflicts with employers' motivations to maintain control over labour process. To overcome these barriers and address inefficiencies in upscaling manufacturing and infrastructure development, the authors highlight a critical need for long-term national workforce development strategies. These strategies should be coordinated and integrated across government portfolios including education, skills, industry, and energy/environment, while remaining grounded in place-based skills and labour markets.

Finally in Chapter 11, economic sociologist Tom Barnes takes up questions of transition through a case study that revisits the closure of Australian automotive manufacturing, as a gauge for how workers and industries may respond to looming industrial transitions such as the decarbonisation of energy systems, manufacturing and agriculture. Using longitudinal data on employment outcomes, Barnes finds that thousands of workers elected to stay in the manufacturing sector despite labour market transition programs that discouraged it, on the basis of limited future opportunities. Barnes' analysis draws on the concept of "occupational citizenship" to characterise this attachment to a sector ostensibly in decline, even with the provision of government retraining and placement programs. The implications for green industrial transitions are twofold: first, in terms of workplace skills Australia has retained significant manufacturing capability, even in light of recent sectoral withdrawals such as automotive manufacturing; and second, there is a need to ensure decarbonisation and transitions policy takes account of not only existing occupational skillsets, but also to less obvious factors including connections to existing occupational communities and labour processes.

When read together, the chapters in this collection raise a set of themes that extend beyond individual contributions and beyond climate change, including the risks of critical skills and labour shortages across multiple domains simultaneously, the capacities of workers to deploy their skills beyond narrow sectoral definitions, and the various ways in which neoliberalism continues to challenge working life. Ultimately, the volume reflects a moment where the challenges of climate change cannot be held apart – empirically or analytically – from the many other dimensions re-shaping working lives. The field of practice we establish here is forthright in its interdisciplinarity and its close connectedness to the daily realities of working life. It comes from an understanding that the immense challenges of future work in a climate-changed future involve a deep and applied rethinking of skills, capacities, and frameworks for understanding labour.

Notes

1 Commonwealth of Australia, *Royal Commission into National Natural Disaster Arrangements Report.*
2 Foley, "Australia Leases US Firebombing Aircraft."
3 Räthzel and Uzzell, *Trade Unions in the Green Economy*; Lipsig-Mumme and McBride, *Work in a Warming World*; Räthzel, Stevis and Uzzell (eds), *Palgrave Handbook of Environmental Labour Studies.*
4 Climate Change Authority, "Sector Pathways Review."
5 Johnson et al., "The Invisible Labour of Climate Adaptation."
6 Parsons and Natarajan, "Geographies of Labour."
7 Pearse and Bryant, "Labour in Transition"; Kleinheisterkamp-Gonzalez, "The Case for an Environmental Labor Geography"; Coe, "Towards a Political Ecology of Labour?"
8 Among the work of so many scholars, key moments for labour in twentieth and twenty-first century industrial transformation have been documented in, for example, Braverman, *Labor and Monopoly Capital*, Massey, *Spatial Divisions of Labour*, Piore and Sabel, *The Second Industrial Divide*, Kassler-Harris, *Out to Work*, Sassen, *The Global City*; Munck, *Globalization and Labour.*
9 Again, a comprehensive list of such contributions is limited by space, but see for example: Acker, "Inequality Regimes,"; McDowell, "The Lives of Others"; Mitchell, Marson and Katz, *Life's Work*; Tronto, *Moral* Boundaries; Weeks, *The Problem With Work*; Adkins and Dever, "Gender and Labour in New Times"; Edwards and Wajcman, *The Politics of Working Life.*
10 Alderslade, "Close Calls."
11 Stein and Carr, *All Hands on Deck.*
12 The *All Hands on Deck* project includes two edited collections: this volume, alongside its sibling, Stein and Carr, *Designing Through Planetary Breakdown: Locating Material Knowledge and Practical Skill.*
13 Abram et al. "Just Transition."
14 Denham and Rickards, *Climate Impacts at Work*; Johnson et al., "The Invisible Labour of Climate Change Adaptation."
15 Mazzochi, "An Answer to the Jobs-Environment Conflict"; Stevis et al., *Just Transitions.*
16 See for example: Abram et al. "Just Transition"; Cha, *A Just Transition for All*; Flanagan and Goods, "Climate Change and Industrial Relations." In "The Just

Transition," Velicu and Barca put forward a persuasive case for why the concept *should* extend beyond the wage relation. Newell and Mulvaney take a broad but critical political economic view in "The Political Economy," which is instructive, focusing on equitable energy access and managing contradictions in addressing energy and climate justice simultaneously, alongside justice for fossil fuel workers.

17 International Labour Organisation, *Guidelines for a Just Transition*.
18 Abram et al. "Just Transition,"; Cha, *A Just Transition for All*; Healy and Barry, "Politicizing Energy Justice."; Heffron and McCauley, "What is the Just Transition?"; Snell, "'Just Transition'?"; Weller, "Just Transition?"
19 Parsons and Natarajan, "Geographies of Labour."
20 Velicu and Barca, "The Just Transition."
21 See for example, Carr, "Maintenance and Repair," and Stein, "Masculinity and Material Culture."
22 Lai et al. "A Place-Based Just Transition Framework"; Eadson et al. "Decarbonising Industry"; Stambe et al., "Jobs Aren't Enough."
23 See also: Carr, "Repair and Care."
24 Snell, "Just Transition"; Banerjee and Schuitema, "How Just Are Just Transition Plans?"; Schuster et al., "The Unjust Just Transition?"
25 Pearse and Bryant, "Labour in Transition."
26 Carr, "Maintenance and Repair"; Pearse and Bryant, "Labour in Transition"; Velicu and Barca, "The Just Transition."
27 Legislated net zero commitments are relatively recent in Global North contexts. The United Kingdom became the first major economy to legislate net zero targets by amending the Climate Change Act 2008 in mid-2019 (Climate Change Act 2008, c. 27, as amended by The Climate Change Act 2008 (2050 Target Amendment) Order 2019, SI 2019/1056). This was followed by Canada and the European Union in 2021.
28 Rickards et al., "Australia's National Climate."
29 Denham et al., "The Jobs of Climate Adaptation."
30 See Lai et al., "A Place-Based Just Transition Framework,", though what we are referring to here extends beyond their approach to incorporate not only adaptation planning, but also intersecting infrastructural needs – housing, education, transport and social services – which are also an essential part of transition planning.
31 Denham et al., "The Jobs of Climate Adaptation."
32 Pearse and Bryant, "Labour in Transition."
33 This attentiveness to our own context might be deemed unnecessary by some, though we note that when publishing in international journals – often based in other Global North countries – Antipodean researchers are commonly asked to justify the 'global' relevance of our work.
34 This was to include a federal Environmental Protection Agency.
35 Climate Council, "State of Queensland."

Bibliography

Abram, Simone, Ed Atkins, Alix Dietzel, Kirsten Jenkins, Lorna Kiamba, Joshua Kirschner, Julia Kreienkamp, Karen Parkhill, Tom Pegram and Lara M. Santos Ayllón. "Just Transition: A Whole-Systems Approach to Decarbonisation." *Climate Policy* 8 (2022): 1033–49.

Acker, Joan. "Inequality Regimes: Gender, Class and Race in Organisations." *Gender & Society* 20, no. 4 (2006): 441–64.

Adkins, Lisa and Maryanne Dever. "Gender and Labour in New Times: An Introduction." *Australian Feminist Studies* 29, no. 79 (2014): 1–11.

Liz, Alderslade. "Close Calls for Aged Care Facilities in Bushfire Zone." *Talking Aged Care*. 17 January 2020. https://www.agedcareguide.com.au/talking-aged-care/close-calls-for-aged-care-facilities-in-bushfire-zones

Banerjee, Aparajita and Geertje Schuitema. "How Just are Just Transition Plans? Perceptions of Decarbonisation and Low-Carbon Energy Transitions Among Peat Workers in Ireland." *Energy Research & Social Science* 88 (2022): 102616.

Braverman, Harry. *Labor and Monopoly Capital: The Degradation of Work in the 20th Century*. Monthly Review Press, 1974.

Carr, Chantel. "Repair and Care: Locating the Work of Climate Crisis." *Dialogues in Human Geography* 13, no. 2 (2022): 221–39.

Carr, Chantel. "Maintenance and Repair Beyond the Perimeter of the Plant: Linking Industrial Labour and the Home." *Transactions of the Institute of British Geographers* 42, no. 4 (2017): 642–54.

Cha, J. Mijin. *A Just Transition for All: Workers and Communities for a Carbon-Free Future*. MIT Press, 2025.

Climate Change Act 2008, c. 27, as amended by The Climate Change Act 2008 (2050 Target Amendment) Order 2019, SI 2019/1056. https://www.legislation.gov.uk/uksi/2019/1056/contents/made

Climate Change Act 2022 (Cth) No. 37, 2022. https://www.legislation.gov.au/Details/C2022A00037

Climate Change Authority. "Sector Pathways Review." 3 October 2024. https://www.climatechangeauthority.gov.au/sector-pathways-review

Climate Council. *State of Queensland: Disaster Ground Zero*. 9 October 2024. https://www.climatecouncil.org.au/resources/state-of-queensland-disaster-ground-zero/

Coe, Neil. "Afterword: Towards a Political Ecology of Labour?" *Area* 53 (2021): 450–53.

Commonwealth of Australia. *Royal Commission into National Natural Disaster Arrangements Report*, 2020.

Copernicus Climate Change Service (C3S). *Global Climate Highlights 2024*. 10 January 2025. https://climate.copernicus.eu/global-climate-highlights-2024

Denham, Todd and Lauren Rickards. *Climate Impacts at Work: Supporting a Climate Ready Workforce*. Climate Resilience Living Lab, RMIT University, 2022.

Eadson, Will, Bregje van Veelen and Stefan Backius. "Decarbonising Industry: A Places-of-Work Agenda." *The Extractive Industries and Society* 15 (2023): 101307.

Edwards, Paul and Judy Wajcman. *The Politics of Working Life*. Oxford University Press, 2005.

Environment Protection and Biodiversity Conservation Act 1999 (Cth) No. 91, 1999. https://www.legislation.gov.au/Details/C2024C00598

Flanagan, Frances and Caleb Goods. "Climate Change and Industrial Relations: Reflections on an Emerging Field." *Journal of Industrial Relations* 64, no. 4 (2022): 479–98.

Foley, Mike. "Australia Leases US Firebombing Aircraft in the Winter. So What Happens when LA Burns in January?" *Sydney Morning Herald*, 9 January, 2025. https://www.smh.com.au/politics/federal/fire-seasons-are-becoming-a-constant-global-threat-this-may-leave-australia-on-its-own-20250109-p5l30j.html

Future Made in Australia Act 2024 (Cth). https://www.legislation.gov.au/Details/C2024A00119

Healy, Noel and John Barry. "Politicizing Energy Justice and Energy System Transitions: Fossil Fuel Divestment and a 'Just Transition.'" *Energy Policy* 108 (2017): 451–9.

Heffron, Raphael and Darren McCauley. "What is the Just Transition?" *Geoforum* 88 (2018): 74–7.

International Labour Organisation. *Guidelines for a Just Transition Towards Environmentally Sustainable Economies and Societies for All.* ILO (2015). https://www.ilo.org/publications/guidelines-just-transition-towards-environmentally-sustainable-economies

Johnson, Leigh, Michael Mikulewicz, Patrick Bigger, Rithodi Chakraborty, Abby Cunniff, Joshua Griffin, Vincent Guermond, Nicole Lambrou, Megan Mills-Novoa, Benjamin Neimark, Sara Nelson, Costanza Rampini, Psang Sherpa and Gregory Simon. "The Invisible Labour of Climate Change Adaptation." *Global Environmental Change* 83 (2023): 102769.

Kessler-Harris, Alice. *Out to Work: A History of Wage-Earning Women in the United States.* Oxford University Press, 1982.

Kleinheisterkamp-Gonzalez, Nicole. "The Case for an Environmental Labor Geography: The Role of Organised Labor in the Climate Crisis." *Progress in Human Geography* 47, no. 2 (2023): 317–32.

Lai, Huei-Ling, Patrick Devine-Wright, Jo Hamilton, Sarah Mander, Diarmaid Clery, Imogen Rattle, Abigail Martin, Stacia Ryder and Peter Taylor. "A Place-Based, Just Transition Framework can Guide Industrial Decarbonisation with a Social Licence." *Energy Research & Social Science* 121 (2025): 103967.

Lipsig Mumme, Carla and Stephen McBride. *Work in a Warming World.* Queens University Press, 2015.

Massey, Doreen. *Spatial Divisions of Labour: Social Structures and the Geography of Production.* Palgrave MacMillan, 1984.

Mazzocchi, Tony. "An Answer to the Jobs-Environment Conflict?" *Green Left Review*, 8 September 1993. https://www.greenleft.org.au/content/answer-jobs-environment-conflict

McDowell, Linda. "The Lives of Others: Body Work, the Production of Difference, and Labor Geographies." *Economic Geography* 91, no. 1: 1–23.

Mitchell, Katharyne, Sallie Marston and Cindi Katz. *Life's Work: Geographies of Social Reproduction.* Wiley-Blackwell, 2004.

Munck, Ronaldo. *Globalisation and Labour: The New "Great Transformation".* Zed Books, 2002.

Newell, Peter and Dustin Mulvaney. "The Political Economy of the 'Just Transition'." *The Geographical Journal* 179, no. 2 (2013): 132–40.

Net Zero Economy Authority Act 2024 (Cth) No. 85, 2024. https://www.legislation.gov.au/Details/C2024A00085

Offshore Electricity Infrastructure Act 2021 (Cth) No. 120, 2021. https://www.legislation.gov.au/Details/C2025C00179

Parsons, Laurie and Nithya Natarajan. "Geographies of Labour in a Changing Climate." *Area* 53, no. 3 (2021): 406–12.

Pearse, Rebecca and Gareth Bryant. "Labour in Transition: A Value-Theoretical Approach to Renewable Energy Labour." *Environment and Planning E: Nature and Space* 5, no. 4 (2021): 1872–94.

Piore, Michael and Charles Sabel. *The Second Industrial Divide: Possibilities for Prosperity.* Basic Books, 1986.

Räthzel, Nora, Dimitris Stevis and David Uzzell. (eds.) *Palgrave Handbook of Environmental Labor Studies.* Springer International Publishing, 2021.

Räthzel, Nora and David Uzzell, eds. 2013. *Trade Unions in the Green Economy: Working for the Environment*. Routledge.

Rickards, Lauren, Timothy Neale and Matthew Kearnes. "Australia's National Climate: Learning to Adapt?" *Geographical Research* 55, no. 4 (2017): 469–76.

Sassen, Saskia. *The Global City: New York, Lonfon, Tokyo*. Princeton University Press, 2001.

Schuster, Antonia, Maksymilian Zoll, Ilona Otto and Franziska Stölzel. "The Unjust Just Transition? Exploring Different Dimensions of Justice in the Lignite Regions of Lusatia, Eastern Greater Poland, and Gorj." *Energy Research & Social Science* 104: 103227.

Snell, Darryn. "'Just Transition'? Conceptual Challenges Meet Stark Reality in a 'Transitioning' Region in Australia." *Globalizations* 15, no. 4 (2018): 550–64.

Stambe, Rose-Marie, Greg Marston, Darryn Snell and Jeremy Moss. "Jobs Aren't Enough: Redefining Just Transitions in Australia with Community Voices." *Energy Research & Social Science* 122 (2025): 103999.

Stein, Jesse Adams. "Masculinity and Material Culture in Technological Transitions in the Workplace: From Letterset to Offset-Lithography, 1960s–1980s." *Technology & Culture* 57, no. 1 (2016): 24–53.

Stein, Jesse Adams and Chantel Carr, eds. *Designing through Planetary Breakdown: Locating Material Knowledge and Practical Skill*. Routledge, 2025.

Stein, Jesse Adams and Chantel Carr. *All Hands on Deck*. Sydney, 2023. https://allhandsondeck2023.com/

Stevis, Dimitris, Edouard Morena and Dunja Krause. "Introduction: The Genealogy and Contemporary Politics of Just Transitions." In *Just Transitions: Social Justice in the Shift Towards a Low-Carbon World*, edited by E. Morena, D. Krause and D. Stevis. Pluto Press, 2020.

Tronto, Joan. *Moral Boundaries: A Political Argument for an Ethic of Care*. Routledge, 1993.

Velicu, Irina and Stefania Barca. "The Just Transition and its Work of Inequality." *Sustainability: Science, Practice and Policy* 17, no. S2 (2020): 263–73.

Weeks, Kathi. *The Problem with Work: Feminism, Marxism, Antiwork Politics and Postwork Imaginaries*. Duke University Press, 2011.

Weller, Sally. "Just Transition? Strategic Framing and the Challenges Facing Coal Dependent Communities." *Environment and Planning C: Politics and Space* 37, no. 2 (2019): 298–316.

PART I

Skills and Training

1

"WHAT'S YOUR APOCALYPSE SKILL?"

Revaluing Flexible Skills for the Apocalyptic Good-Life

Kezia Barker and Rosie Cox

Introduction

In 2020 the notion of the "frontline" protecting nations from "dangerous" external threats expanded beyond the military, becoming diversified and geographically fragmented. Medical staff were applauded as they battled to save Covid-19 patients filling hospital wards, but workers keeping supermarket shelves stocked with food and cities cleared of rubbish were also regularly referred to as operating on the Covid-19 "frontline" and as vital to their nation's efforts. This demonstrates how crisis can, albeit temporarily and by no means in economically redistributive ways, make visible and disrupt the normal attribution of value that attaches to different occupations and associated skills. This revelatory and re-valuatory function of crisis extends to anticipatory evocations. "Preppers" are a sub-cultural group who imaginatively explore different crisis scenarios to hone their preparations, including stockpiling food and equipment and developing survival skills.[1] By bringing imagined catastrophe into the present and everyday, prepper scenarios wrench people, places and things out of their existing forms of signification and consider them anew. This imagining not only offers opportunities for better preparedness for crisis scenarios but also opens space to re-examine and revalue current ways of being that might bring about those same crises.

In an online post within a closed UK social media prepping group in November 2020, a member asks: "How is your job useful to any major SHTF[2] situation? If not, are you thinking of changing your job?" Revealing themselves to provide night-time security at a metal recycling factory, they comment, "Mine's useful because of the security, but I also have access to some heavy duty machinery and vehicles." Across over 250 responses, often with typical

DOI: 10.4324/9781003512240-3

British self-deprecating humour, contributors lifted their everyday work out of its contemporary bundling of tasks and value and imagined themselves in apocalyptic futures. The utility of manual jobs would be enduring, as one dry response exemplified: "Carpenter. Same shit, different day." Responses jokingly speculated what exactly they'd need from a carpenter in a zombie apocalypse: "yeah I was thinking more like some decking, maybe a water feature?" Overall, the responses reference the haptic, hands-on skills Chantel Carr and Chris Gibson refer to as typically undervalued.[3] No one identifies as a lawyer, banker or politician, although one brand designer points to their further skills as a home-brewer: "No one will be wanting a new logo, but maybe the booze will win me friends." The hairdresser who laments their uselessness is reassured by other members, citing experience during and after successive Covid-19 lockdowns: "you'll be in demand!" As long, she is told, as she gets some battery-powered clippers and spare batteries: she'll need to be able to operate off-grid. Other group members comment on each other's responses, musing and challenging notions of what will be useful in this undefined but dark future: "I'm a carer. Is that useful?" received responses such as "without a doubt tbf," "yes very!" and "you're the next best thing if we don't have doctors and nurses!"

The question this online group was contemplating can be placed within a genre of cultural scenario analysis we loosely categorise as *What's your apocalypse skill?* – a perennial discussion point in prepping communities.[4] In this example, current world occupations are reflected on, judged and revalued through a non-specific ("major SHTF") catastrophic future scenario. Other examples are framed around the survival skills and capabilities a person would bring to catastrophic or post-apocalyptic worlds. Sometimes, the question is directly anchored around the skills needed for an ideal survival unit or "MAG" (mutual assistance group); or skills that may be traded. Overall, the question *what's your apocalypse skill?* seeks to lift a person's occupation and skills out of their contemporary value hierarchy, and reattribute value based on the speculative coordinates of a new world. Lesley Head attends to how knowledges, technologies and practices (and humans-nonhumans) have been "bundled together" differently in given locales leading to regionally distinctive Holocenic collectives.[5] The question these apocalyptic-skill musings seem to be playfully probing is how disrupted Anthropocenic futures will – as human impact on our planet defines a new geological epoch – untie, reassemble and revalue these bundles of knowledge-technology-practice. The scenarios contemplated and those that drive prepping are not necessarily about climate changed-futures. The effects of imposed austerity, the everyday impacts of economic and health hardships and the withdrawal of the welfare state, all feature as drivers and justifications for prepping.[6] However, their imaginative engagement with apocalypse is all too resonant with climate-disrupted future-presents.

In this chapter we reflect on this genre of scenario analysis as it emerges in Kezia Barker's ongoing ethnography of UK prepper groups, to explore the motivations and implications of the re-valuation of practical capabilities as survival skills for catastrophic futures. In the section "Valuation and flexibility at work in late modernity, the apocalypse, and beyond", we review work in feminist and economic geography on skill valuation and concepts of flexibility, and consider how forms of crisis, catastrophe and apocalypse disrupt or unveil true meaning and value. In the section '"What's my apocalypse skill? … Depends on the apocalypse." – Flexibility as a relational, embedded orientation to uncertainty', we use empirical examples to explore how preppers value the adaptable skills gained through practical occupations, enabling them to recognise potential in both their material resources and their own capabilities. What comes to be valued most highly, we argue, are specific forms of adaptability centred on the capacity of the skilled individual to take advantage of a changing environment. In Conclusion, we argue that imaginations of apocalypse and conditions of postapocalyptic survival disrupt capitalist valuation systems by bringing commodities, people and skills into regimes of valorisation that are not centred on accumulation. This is a shift in emphasis from questions of *how much do I get?* to *how long can I live on for?* In exploring scenarios of apocalyptic futures, the prepper gestures beyond the neoliberal present by imagining and preparing for alternatives in ways that centre practical, creative, embodied flexibility.

Valuation and flexibility at work in late modernity, the apocalypse, and beyond

Research within economic and feminist geography has extensively highlighted and challenged the uneven economic and cultural de/valuation of different forms of work, particularly physical skilled labour and caring roles.[7] Guéry and Deleule identified explicit efforts within early capitalist processes to reorganise modes of production to create, fragment and thus extract maximum value from physical labour, separating "mind" tasks from "hands on" manual tasks.[8] This was entangled with the economic and cultural de/valuing of physical labour, reflected in both economic recompense paid to workers and the perceived cultural capital of different occupations. This devaluation of manual tasks extends to the individuals undertaking that work, with those connected to maintenance and repair suffering further, through discursive connection to feminised care-based tasks.[9]

In addition to the (over)valuing of cognitive work, late modern labour has been characterised by an increased value being put on certain kinds of flexibility at work.[10] In the gig economy, flexibility shapes an ever larger numbers of jobs. As Doreen Massey and colleagues highlighted, however, there is a profound difference in what it means to be flexible for highly-paid and lower paid workers, with gendered implications of demands for ultimate flexibility from "zero drag" workers.[11] The fluctuation of perceived future skill requirements brings with it the additional threat of the expendability of whole professions. Workers are entreated to be ready to retrain and redeploy, "transferable skills" are decoupled from their association with particular jobs, and we are at continual risk of becoming under or over-qualified for the market.[12]

The downward trajectory of a present always-already out-of-joint with the demands of the future is now met with the hallmark of the Anthropocene: fears for environmental catastrophe. If the Holocene allowed enough stability for the flourishing and complexification of civilisation, and if capitalism created, not just devalued, manual labour,[13] what does the instability of the Anthropocene demand? In particular, what forms of flexibility? And what will happen to those jobs, skills and people only suited to the Holocene's defunct operational requirements?[14] Over and above the destabilisation of values under conditions of the slow crisis of precarity,[15] the presence and spectre of environmental catastrophe pushes value, meanings and utility of things into apocalyptic flux.

Tracing the etymological origins of eventful disruptions, Rebecca Solnit identifies emergency as a separation from the familiar, a sudden emergence into a new atmosphere; and catastrophe as "an upset of what is expected … a plot twist."[16] The next stage in this disruption of the Real is apocalypse. In its original Greek translation, apocalypse is an "unveiling," a defining revelation of previously unknown "truths" that will fix the true value of things and their appearance in space. Eva Horn writes: "What is revealed by the apocalypse is the true value and the power of everything and everyone. The end of the world is the unmasking of all things, the manifestation of their true essence."[17]

This capacity of apocalypse to reappraise the true value and essence of "everything and everyone" finds expression in cultural fascination for apocalyptic scenarios. If you had to leave your home in an emergency, what is the one thing you would save? If you were shipwrecked on a desert island, what one item would you take? In facing catastrophic futures, what knowledge will be vital, to survive or reboot society? These concerns have spawned a genre of books cataloguing essential knowledge (from encyclopaedias, to James Lovelock's *The Earth and I*,[18] to *The Knowledge*[19]), alongside bunkers and doomsday vaults seeking to transport archives of seeds and equipment as our travelling companions into unknown futures.[20] Apocalypse as the unveiling of true value is also extensively explored through survivalist TV, apocalyptic and post-apocalyptic film. As Nicholas Beuret and Gareth Brown write in an analysis of *The Walking Dead*:

> It is a world saturated with waste and ruins – with objects severed from their previous use values. Abandoned factories, empty buildings, quiet roads. More than this, social roles no longer hold their value. It is not who we are that forms the basis of action, but what our bodies can do.[21]

In contrast to the rigidity of pre-existing social roles, through apocalypse imaginings bodies and things are flexibly reappraised as they meet with different survival challenges.

The question remains, however, of exactly which social roles are being cast aside in this return to the primacy of what "bodies can do." In an erudite analysis of US survivalist tv shows including National Geographic's long-running series *Doomsday Preppers*, Casey Kelly argues that the crisis motif in US popular culture entrenches an apocalyptic version of hegemonic manhood through the recuperation of traditional manly skillsets as survival tools.[22] Apocalypse functions here to unveil the consequences of a feminised dependency culture, constructing manly skills as "necessary for collective survival."[23] Josef Nguyen builds on this analysis, arguing that doomsday prepper discourse is converging with US maker culture in ways that reinforces the "racial politics of craft and DIY movements as anxious responses to modern industrialisation in the US."[24] This produces an idealised DIY subjecthood, framing skills of making and crafting as necessary in preparing for future apocalypse. By "recuperating" what are characterised as basic, primitive, or premodern survival skills "lost" to the modern, urban, white US citizen, a racialised development hierarchy constructs other geographic settings and non-white Others as left behind and always-already apocalyptic.

These critical accounts do vital work in linking crisis anticipations to perceptions of what is thought to be undervalued or veiled in everyday modern life: a crisis of masculinity and loss of white male privilege. These insights are applicable to places and sub-cultures on the spectrum of prepping communities; in certain national contexts they may be the dominant expression. However, there is a tendency for accounts of prepping to collapse all subcultures into a stereotypical US version. From US roots,[25] prepping has become a global phenomenon, with national prepping cultures in France,[26] Sweden,[27] the UK[28] and South Africa[29] subject to academic analysis. This fledgling research field is unpicking the influence of varied cultural histories, political identities and experiences of crisis on the spectrum of prepping subjectivities and practices, identifying varied manifestations and inscriptions of individualism and resourcefulness. Kelly[30] and Nguyen[31] both draw exclusively on and conflate media representations of preppers with community self-representations. This perpetuates the media pathologisation of prepping and dismissal of individual and community self-perceptions of crisis vulnerability, and misses nuances of "self-reliance" as a concept that performs in different ways across political fields.[32] This underlies the importance of understanding prepping cultures through ethnographic research.

"What's my apocalypse skill? ... Depends on the apocalypse." – Flexibility as a relational, embedded orientation to uncertainty

Survival skills can be defined around a core of practices that sustain autonomous life. The survivor is disembedded from the relations and infrastructures of contemporary life and required to draw on resources and affordances in environments and situations of uncertainty, unfamiliarity and scarcity.[33] Emphasis is placed on the necessity of existing off-grid, beyond the operational requirements of a pre-determined "techno-environmental system,"[34] in scenarios of infrastructural separation, disruption or collapse. Survival disrupts any easy distinction between production and reproduction. Care and social reproduction work are lifted away from an unremunerated, unskilled support role in the private sphere to become the sole focus of life – the basis of survival.

A plethora of survival books, schools offering courses and shared online resources detail the specificity of these skills, focused on the requirements for sustaining metabolic individualism. In a social media post from a large UK survival group,[35] survival is defined as:

> Survival is all about being prepared and equipped with the essential skills to thrive in any situation. ... Honing these skills will give you the confidence to tackle challenges head-on. Here are the top 5 survival skills to learn in the UK:
>
> - Fire-starting
> - Shelter building
> - Navigation
> - Foraging
> - First aid

This list, perhaps purposely designed to cause contention and traction, solicited alternative ideas for the top five slot, with responses adding water purification skills; knife skills and green woodworking; trapping and game prep/outdoor cooking; camp craft and cordage. However, it is the notion of being "prepared" (a hyper-affective, anticipatory awareness of multiple potential futures) and "equipped" (how these skills attach to the individual) with these essential skills to thrive in "any situation" (flexibility to uncertainty) that is key to understanding how the *what's your apocalypse skill?* question disrupts and reattributes the value of skills. Following from the debates reviewed above, we argue that flexibility takes on a specific form across prepper cultures. In the following we attend to three key attributes: flexibility as a material possibility; activated by skilled embodied practice; and applied to environmental affordances.

First, flexibility is a quality that can be materialised. Objects or materials that are physically as well as imaginatively flexible and adaptive to different survival contexts are highly prized. Objects may be designed to be multi-functional or simply put to multiple possible uses. The night-time security officer, in the discussion thread that opened our chapter, visualises potential in the machinery and the metal he has access to. He imagines survival scenarios generating use-value in worlds beyond its recirculation in the capitalist economy (with sustainability credentials attached).

The visualisation of potential in materials is also appreciated in studies of maker cultures, as part of the reappraisal of the "head" work of skilled practice.[36] It is not just different uses or applications of material things that is significant, however, but the multiple and flexible values layered upon them. What could a thing become if embedded in a different set of relations, decoupled from usual assemblages, or part of a different legal and moral universe? With the anticipated infrastructural and social collapse that is gestured to in the *what's your apocalypse skill* question, entities are detached from their prescribed place in the world and float freely, with the possibility of being rescripted and re-assigned different identities, placed into different assemblages. In these scenarios, present value detaches from objects as the existing cultural, legal and physical apparatus for valuing them collapses.[37] Value is imaginatively reattached through the specificity of assumptions about future scenarios and the moral and legal worlds they trace.[38] By finding value in mundane objects and centring the significance of "off-grid" materialities, preppers in small ways are challenging the interface of infrastructure, power and value of/over life.[39]

Second, material potentiality is recognised and activated through skilled practice, with value placed on the capacity to tinker, experiment and be a "handy" problem solver. Carr and Gibson blur the head/hand divide when they write of makers being "equipped with the sensibilities and disposition to conceive of things-at-hand as only ever temporary gatherings of matter and idea, which can disperse and be reassembled elsewhere in new combinations."[40] Material potentiality is therefore not something simply built into objects (like the fabled Swiss army knife) but is relational and performative, needing skill and vision to identify, extract, activate and channel its value. Material objects become a manifestation of the skill they demand: "to learn or discover them as manual, physical, embodied, personal skill."[41] Equally, skill is not simply brought to bear on innate materials but is demonstrated through a considered negotiation with material liveliness.[42] This emerges in a reverence for practice, use and training, and an openness to experimental informal learning. Interviewees described how they moved between online and offline learning resources, and from learning to testing, as they acquired survival skills; in this example, one resource for fire-starting:

Interview A: Somebody's mentioned on, em, some page or other that I've been reading, about fatwood, fatwood, and I'm, 'What's all that about?' So, I go on YouTube, fatwood, what's that about? [So, I had to learn] what it was, where I can find it, how to obtain it, how to use it. So, the last time I went out, I went out and used it. Brilliant! Absolutely brilliant.

Another emphasised the importance of putting online knowledge to the test, through embodied practice:

Interview B: Em, most of it, most tends to be online … I do a lot of learning myself, as in reading books and stuff, and going out and practising skills and stuff like that … I mean, it's one thing reading, but you've got to kind of go out and you've got to try to do it, you know what I mean?

The possession of skill is never a fixed condition, but an ongoing practice of learning, as the SAS guide expounds: "skills and experience must not be allowed to get rusty and you must extend your knowledge all the time."[43] You need to keep your skills, not just your knife, sharp and ready for use. New skills may have to be learnt to fill the gaps left exposed by uncertain futures:

Interview F: Am I as knowledgeable enough as I would need to be? No, no way. But a lot of these … I am trying to learn new skills new things that I think well, that might be useful … I have had pigs … I thought, well if I'm going to go continue to be a meat eater, I ought to know the whole process … Well, I'm more informed than I was. You know, it's a new skill. I wouldn't, I wouldn't starve. But I wouldn't ever go into any competitions to do the butchery of an animal again. But I have learnt a lot more than I knew before.

At its extension, what this imagines is a body operating in isolation from the baggage of modern life, replacing technology with technique: "The more you know, the less you carry."[44] This is expressed explicitly through shows such as *Naked and Afraid*, which strips contestants and places them in survival situations. In answer to a moderator's question in an online prepper group "What would be your job in post-apocalyptic society?," one respondent comments: "I'm glad we're discussing skills and not our toys, you can lose everything, but skills you take with you."

The SAS survival guide recommends: "Any equipment you have must be considered a bonus. Lack of equipment should not mean that you are unequipped, for you will carry skills and experience with you."[45] Here, it is skill that "equips" the survivor, as a "carry" from which one cannot be separated. It is significant that despite the imaginative openness of *what's your apocalypse skill?* scenario musings, contributors rarely imagine themselves anew, as is typical in other online communities such as *Second Life*. Instead, the bundling of person and skill that form the basic building blocks of existing occupations are carried into alternative, uncertain futures. Whether this then reproduces or disrupts patriarchal codings of traditional manly skills depends on the culture of the sub-group. Skills are revalued (for example, care/reproductive labour is scripted as vital to survival), but existing stereotyped attachment between bodies and skills can remain.

Third, material potentiality and embodied skill require ongoing adaptation and sensitivity to unstable environments, through the spectrum of crisis scenarios preppers encounter and, crucially, imagine. Contextual improvisation and perceptual monitoring has been identified by Tim Ingold as significant during the unfolding of a task: "the capacity to respond to unfolding iterations with materials, to use slowly accrued haptic knowledge to manipulate processes on the fly, and to judge how to counteract error and seize opportunities as they evolve."[46] An unexpected knot in wood whilst carving a spoon, the adjustments needed to homemade rocket stoves whilst cooking in the woods, or skills as a car mechanic put to use in imaginative survival scenarios, all require "variation in response to the conditions at hand."[47] Skill is therefore constantly challenged by unfamiliar situations both in practice, preparation and in scenarios. Survival is based upon the capacity to anticipate and so have benevolent materials to hand, coupled with prior development and ongoing practice of skills, in order to "activate" affordances that might emerge in uncertain, changing environments.[48] As material potentiality, skill capabilities and shifting environmental conditions relationally produce ecological affordances, what survival opportunities an environment and situation might afford one person, it may not afford another.[49] As one online commentator argued, this reinforces the value of adaptability in the face of the multitude of different imagined and experienced forms of crisis: "What's your apocalypse skill? Kind of depends on the apocalypse really … [that']s why adaptability is your best aim … Jack of all trades is the way to go."

In imaginative iterations of post-apocalyptic, ruined landscapes, this flexible capacity to adapt and respond to environmental affordances often entails picking through and salvaging the "detritus" of the Anthropocene.[50] For Beuret and Brown, the ongoing, irreversible disaster of the Anthropocene necessitates approaching maintenance, repair and care "without hope of renewal," transforming them from practices of restoration to a ceaseless labour of salvage.[51] Stephanie Wakefield however, challenges the imposed victimhood of ruins theory "as if survivors simply aim to survive its ruins or disasters, as if liberal life is all there will ever or could ever be."[52] She argues instead that those living through catastrophe or its aftermath be reconceived as "experimenters" who "weave traditions and tools into new arrangements that suit them."[53] This is echoed by Tatiana Thieme's provocation to think about things "beyond repair": "beyond the validation of acts of maintenance and repair which are associated with the promise of some return to utility – conceptualising breakdown as a modality of rupture, a new phase."[54] This takes us back to notions of apocalypse as openings to alternative futures. Crucially, this opening is grasped through preppers' imaginative scenarios that are used to explore, reshuffle and experiment with notions of essential value and meaningful practice. While apocalypse is understood to unveil the "true value" of people and things, this true value is not one fixed attribute but flexibility and potentiality itself. Preppers emerge as skilled experimenters and imaginative inventors making use of the affordances of apocalyptic scenarios to "rupture" capitalist valuations of their skills, and fundamentally, of themselves.

Conclusion

As Solnit argues, "if paradise now arrives in hell, it's because in the suspension of the usual order and the failure of most systems, we are free to live and act another way.[55] The revaluation and disruption of meaning through catastrophe can, Solnit notes, engender "positive emotions" that "demonstrate that social ties and meaningful work are deeply desired, readily improvised, and intensely rewarding."[56] She identifies the latent qualities of "meaningful work" revealed by catastrophe as "participation, agency, purposefulness, and freedom," as catastrophe "return[s] their sufferers to public and collective life."[57] This desire for meaning and value, agency and purpose is revealed within preppers' survival scenarios. Richard Mitchell, in the genre-defining ethnography of US survivalism *Dancing at Armageddon*, argues that preparedness should be seen as an expression of the desire for unstructured potential, for alternatives, for creative roles in determining value and culture.[58] Preppers are not often those "makers" explored by academia who typically participate in creative processes[59]; rather, they are those still tied up in neo-liberal industrial processes. Yet they see themselves and their skills as something that could be extracted from the restrictive coordinates of the present to be reapplied, and revalued, in another world. In some cases, this contemporary devaluation is seen as a matter of life or death, with prepping justified as a way to avoid dependency on emergency responders or welfare agencies who have categorised certain communities and individuals as unworthy of saving.[60]

Ingold challenges the binary separation of mind/body and the supposed repetitive, physical drudgery of labour, by reframing understandings of technical skill and material knowledge as constituting *deep* concentration. Preppers extend this challenge by taking it deeper still. Just as the survival literature refers to a "survival mindset," "survival mentality" and certain "inner qualities" that interface with survival skills, preppers reframe understandings of technical skill and material knowledge as constituting a person's inherent self and worth. These skills are distilled as those of being practical, flexible and crucially, imaginative.

Notes

1 See Barker, "How to Survive"; Bounds, *Bracing for the Apocalypse*; Campbell, Sinclair and Browne, "Preparing for a World Without Markets"; Kerrane et al., "(Invisible) Displays of Survivalist Intensive Motherhood"; Mills, "Preparing for the Unknown"; Mills, "Obamageddon."
2 Acronym for "Shit hits the fan" – an unspecific crisis or catastrophic situation.
3 Carr and Gibson, "Geographies of Making."
4 We have noticed how this genre of questions spills into other online spaces e.g. Facebook groups supporting home learning during Covid-19, asking, "what skills do you think our children actually need to learn for the future?"
5 Head, *Hope and Grief in the Anthropocene.*
6 Barker, "How to Survive;" Barker, "Awakening from the Sleep-walking Society."
7 England and Lawson, "Feminist Analyses of Work"; McDowell, "The Lives of Others."
8 Guéry and Deleule, cited in Carr and Gibson, "Geographies of Making."
9 Carr, "Maintenance and Repair"; Graham and Thrift, "Out of Order."
10 Standing. The Precariat.
11 Henry and Massey, "Competitive Time-space"; Massey, *Space, Place and Gender*. Zero drag workers include those without any caring responsibilities that might be prioritised above the needs of the organisation.
12 Fraser, *Cannibal Capitalism.*
13 Carr and Gibson, "Geographies of Making."
14 Fenton, "'Bushcraft' and 'Indigenous Knowledge,'" 260.
15 Berlant, *Cruel Optimism.*
16 Solnit, *A Paradise Built in Hell*, 7.
17 Horn, *The Future as Catastrophe*, 22.
18 Lovelock, *The Earth and I.*
19 Dartnell, *The Knowledge.*
20 See Wakefield, *Anthropocene Back Loop*, for a discussion on the "Global Village Construction Set," a civilisation starter kit.
21 Bueret and Brown, "The Walking Dead," 7.
22 Kelly, "The Man-pocalypse.".
23 Kelly, "The Man-pocalypse," 96.
24 Nguyen, "How Makers and Preppers Converge."
25 Mills, "Preparing for the Unknown"; Mills, "Obamageddon."
26 Vidal, "When the Nuclear Winter Came."
27 Rahm, "Who will Survive?"
28 Barker, "How to Survive"; Barker, "Awakening from the Sleep-walking Society."
29 Senekal, "An Ark without a Flood."
30 Kelly, "The Man-pocalypse."
31 Nguyen, "How Makers and Preppers Converge."
32 See in the US context Bounds, "Bracing for the Apocalypse" on black New York preppers responding to the abandonment of racialised communities by FEMA in disasters.
33 Fenton, "'Bushcraft' and 'Indigenous Knowledge.'"

34 Ingold, *The Perception of the Environment*, 289.
35 UKSN Blog.
36 Carr and Gibson, "Geographies of Making."
37 Carr and Gibson, "Geographies of Making."
38 Kabel and Chmidling, "Disaster Prepper."
39 Wakefield, *Anthropocene Back Loop*.
40 Carr and Gibson, "Geographies of Making," 18; emphasis added.
41 Fenton, "'Bushcraft' and 'Indigenous Knowledge,'" 252.
42 Ingold, *The Perception of the Environment*.
43 Wiseman, *SAS Survival Handbook*, 13.
44 Quote often attributed to bushcraft and wilderness survival expert Mors Kochanski.
45 Wiseman, *SAS Survival Handbook*, 13.
46 Ingold, "Five Questions of Skill," 160.
47 Ingold, "Five Questions of Skill," 160.
48 Fenton, "'Bushcraft' and 'Indigenous Knowledge,'" 35.
49 Fenton, "'Bushcraft' and 'Indigenous Knowledge,'" 35.
50 Beuret and Brown, "The Walking Dead," 5.
51 Beuret and Brown, "The Walking Dead," 7.
52 Wakefield, *SAS Survival Handbook*, 104.
53 Wakefield, *SAS Survival Handbook*, 93.
54 Thieme, "Beyond Repair."
55 Solnit, *A Paradise Built in Hell*, 7.
56 Solnit, *A Paradise Built in Hell*, 7.
57 Solnit, *A Paradise Built in Hell*, 7.
58 Mitchell, *Dancing at Armageddon*.
59 Carr and Gibson, "Geographies of Making."
60 Bounds, *Bracing for Apocalypse*.

Bibliography

Barker, Kezia. "Awakening From the Sleep-Walking Society: Crisis, Detachment and the Real in Prepper Awakening Narratives." *Environment and Planning D: Society and Space*, 40, no. 5 (2022): 805–23.

Barker, Kezia. "How to Survive the End of the Future: Preppers, Pathology, and the Everyday Crisis of Insecurity." *Transactions of the Institute of British Geographers* 45 (2020): 483–96.

Berlant, Lauren. *Cruel Optimism*. Durham: Duke University Press, 2011.

Beuret, Nicholas and Gareth Brown. "The Walking Dead: The Anthropocene as a Ruined Earth." *Science as Culture* 26, no. 3 (2017): 330–54.

Bounds, Anna Maria. *Bracing for the Apocalypse: An Ethnographic Study of New York's 'Prepper' Subculture*. London: Routledge, 2020.

Campbell, Norah, Sinclair, Gary and Sarah Browne. "Preparing for a World Without Markets: Legitimising Strategies of Preppers." *Journal of Marketing Management*, *35* (2019): 798–817.

Carr, Chantel. "Maintenance and Repair Beyond the Perimeter of the Plant: Linking Industrial Labour and the Home." *Transactions of the Institute of British Geographers*, 42 (2017): 642–54.

Carr, Chantel and Chris Gibson. "Geographies of Making: Rethinking Materials and Skills for Volatile Futures." *Progress in Human Geography* 40, no. 3 (2016): 297–315.

Clinton, Gregory Stephen Wells. "The Architecture of Safety: Bunker Mentalities and the Construction of Safe Space in America." PhD dissertation. Stony Brook University, 2017.

Dartnell, Lewis. *The Knowledge: How to Rebuild our World from Scratch*. London: The Bodley Head, 2014.

England, Kim and Valerie Lawson. "Feminist Analyses of Work: Rethinking the Boundaries, Gendering and Spatiality of Work." In *A Companion to Feminist Geography*, edited by Lise Nelson and Joni Seager, 77–92. Malden, MA: Blackwell Publishing, 2005.

Fenton, Lisa. "'Bushcraft' and 'Indigenous Knowledge': Transformations of a Concept in the Modern World." PhD dissertation. University of Kent, 2016.

Fraser, Nancy. *Cannibal Capitalism: How Our System is Devouring Democracy, Care, and the Planet and What We Can Do About It*. London: Verso Books, 2022.

Graham, Mark and Nigel Thrift. "Out of Order: Understanding Repair and Maintenance." *Theory, Culture and Society* 24, no. 3 (2007): 1–25.

Guéry, Francois and Didier Deleule. *The Productive Body*, trans. Philip Barnard and Stephen Shapiro. Abingdon: Zero Books, 2014 [1972].

Head, Lesley. *Hope and Grief in the Anthropocene. Re-conceptualising Human-Nature Relations*. Abingdon: Routledge, 2016.

Henry, Nick and Doreen Massey. "Competitive Time-Space in High Technology." *Geoforum* 26, no. 1 (1995): 49–64.

Horn, Eva. *The Future as Catastrophe*. New York: Columbia University Press, 2014.

Ingold, Tim. "Five Questions of Skill." *Cultural Geographies* 25, no. 1 (2018): 159–63.

Ingold, Tim . *The Perception of the Environment: Essays in Livelihood, Dwelling and Skill*. London: Psychology Press, 2000.

Kabel, Allison and Catherine Chmidling. "Disaster Prepper: Health, Identity, and American Survivalist Culture." *Human Organisation* 73 (2014): 258–66.

Kelly, Casey Ryan. "The Man-pocalypse: Doomsday Preppers and the Rituals of Apocalyptic Manhood." *Text and Performance Quarterly* 36, no. 2–3 (2016): 95–114.

Kerrane, Ben, Katy Kerrane, Shona Bettany and David Rowe. "(Invisible) Displays of Survivalist Intensive Motherhood among UK Brexit Preppers." *Sociology* 55, no. 6 (2021): 1151–168.

Lovelock, James. *The Earth and I*. Cologne: Taschen, 2016.

Massey, Doreen. *Space, Place and Gender*. Minneapolis: University of Minnesota Press, 1994.

McDowell, Linda. "The Lives of Others: Body Work, the Production of Difference and Labour Geographies." *Economic Geography* 91, no. 1 (2015): 1–23.

Mills, Michael F. "Obamageddon: Fear, the Far Right, and the Rise of "Doomsday" Prepping in Obama's America." *Journal of American Studies* 55, no. 2 (2019): 336–65.

Mills, Michael F. "Preparing for the Unknown … Unknowns: 'Doomsday' Prepping and Disaster Risk Anxiety in the United States." *Journal of Risk Research* 22, no. 10 (2018): 1267–79.

Mitchell, Richard G. *Dancing at Armageddon; Survivalism and Chaos in Modern Times*. London: University of Chicago Press, 2002.

Nguyen, Josef. "How Makers and Preppers Converge in Premodern and Post-Apocalyptic Ruin." *Lateral* 7, no. 2 (2018): 1–19.

Rahm, Lina. "Who will Survive? On Bodies and Boundaries After the Apocalypse." *Gender Forum* 45 (2013): 1–3.

Senekal, Burgert A. "An Ark Without a Flood: White South Africans' Preparations for the End of White-Ruled South Africa." *Journal for Contemporary History* 39 (2014): 177–95.

Solnit, Rebecca. *A Paradise Built in Hell: The Extraordinary Communities That Arise in Disasters*. New York: Viking, 2009.

Standing, Guy. *The Precariat: The New Dangerous Class*. London: Bloomsbury, 2011.

Thieme, Tatiana A. "Beyond Repair: Staying with Breakdown at the Interstices." *Environment and Planning D: Society and Space* 39, no. 6 (2021): 1092–110.

UKSN Blog. "UKSN Top 5 Survival Skills to Master in the UK," June 13 2024. https://www.uksn.org.uk/post/uksn-top-5-survival-skills-to-master-in-the-uk

Vidal, Bertrand. "When the Nuclear Winter Came: The Food Imagination of the Sub-Culture Prepper." *ESSACHESS – Journal for Communication Studies* 8, no. 2 (2015): 195–204.

Wakefield, Stephanie. *Anthropocene Back Loop Experimentation in Unsafe Operating Space*, London: Open Humanities Press, 2020.

Wakefield, Stephanie. "Infrastructures of Liberal Life: From Modernity and Progress to Resilience and Ruins." *Geography Compass* 12 (2018): e12377.

Wiseman, John. *SAS Survival Handbook: How to Survive in the Wild, in any Climate, on Land or at Sea*. Glasgow: HarperCollins, 2004.

2

LOCALISING THE OPERATION OF GRID-TIED MICROGRIDS FOR FUTURE RESILIENCE

Kathryn Lucas-Healey and Hedda Ransan-Cooper

Introduction

The Black Summer fires of 2019–2020 – which actually began in the Australian *winter* of 2019 – burnt out 24 million hectares of land across every state and territory of Australia, destroying 3,000 homes and causing over AUD$10 billion in damage. Nationally, 33 people died (not including those who died due to poor air quality), and nearly three billion animals were killed or displaced.[1] On the South Coast of New South Wales, the bushfires damaged electricity transmission and distribution networks, causing major outages. Difficulties accessing the affected areas added to the time it took to return power to the region[2] and the situation was intensified by the many tourists visiting the region at the time. Power outages during disasters not only cause disruptions and economic losses; they have compounding effects on other services on which people rely, such as telecommunications, emergency warnings and evacuation centres, water supplies, fuel supplies, heating and cooling, and access to banking.[3] During the Black Summer fires, a shortage of diesel generators, coupled with difficulties accessing liquid fuels, left many communities having to cope without essential supplies and capabilities. Only months later, residents of the same region faced multiple incidents of serious flooding, which again disrupted power supplies. Since these events, local community groups have been considering how they might cope better during climate change-related severe weather events. The ability to keep power running during disasters is seen as key to greater resilience in the region. Localised, grid-tied energy microgrids have been proposed as a novel way to keep the power on when it is most needed.

Grid-tied microgrids are sections of the electricity grid that can be isolated ("islanded") from the main grid to maintain an independent supply to several

DOI: 10.4324/9781003512240-4

properties. Most of the time, the microgrid infrastructure would be connected to and operate as a normal part of the grid. These types of microgrids use the existing poles and wires of the electricity distribution network, and include local generation and storage technologies such as batteries, solar panels or diesel generators. They can be distinguished from "stand-alone power systems," which are microgrids that are never connected to the main grid. In a grid-tied microgrid there are two main possibilities for locating generation and storage infrastructure: it can be owned by residents and located on private property *behind* their electricity meters; or it can be connected *in front of the meter* on the network (or street) side with some other kind of ownership model. Currently, when there is a power outage, most private "behind-the-meter" solar and battery systems are switched off, because they are not designed to operate safely in islanded mode. After experiencing a major outage such as those following the Black Summer fires and east coast flooding events, it is unsurprising that someone might investigate using their own solar panels or batteries to maintain supply to their property. If that could be done, then why not also supply power to their neighbours? A microgrid is an extension of this logic.

In this chapter we connect key aspirations for localised, grid-tied microgrids with what is needed to realise them, focusing on a case study located in the Eurobodalla Shire on the South Coast of NSW, Australia. The Eurobodalla Shire sits in a mountainous coastal area 300 kilometres south of Sydney and 150 kilometres east of the national capital of Canberra. National parks geographically bound the region to a strip of coastline characterised by many inlets and bays. There are three large towns: the administrative centre of Moruya, and the commercial centres of Batemans Bay and Narooma, with many smaller towns and localities scattered along the coast. The geography means that some towns are serviced with electricity via a single spur line. Spur lines can constitute a single point of failure that causes an outage, but on the other hand, they might make a community easily "islandable." An estimated 31.5 percent of dwellings in the Shire have solar panels installed, which is slightly above the NSW average.[4] Tourism and transition to retirement are major parts of the local economy and result in a large transient population. Due to the Shire's proximity to the cities of Sydney and Canberra, a considerable proportion of housing consists of seasonally occupied holiday homes. Prior to the Black Summer and Covid-19 pandemic, over 30 percent of housing stock in the Eurobodalla was usually vacant.[5] These factors result in energy consumption patterns that are fluctuating and difficult to predict, making a microgrid especially complex to operate while islanded, because it must be able to match supply and demand. The combination of physical, human and infrastructural geographies and needs make the Eurobodalla a unique design, governance and operational challenge, as indeed would be the case for any candidate site.

In the analysis that follows we draw on empirical qualitative data from 19 interviews with professionals working in energy policy, network operation, community groups, research, local government and energy market bodies. Beginning with a critical overview of why microgrids are seen as a potential solution to some of the challenges Eurobodalla Shire communities face, we recount themes of energy security and sovereignty, decarbonisation imperatives, overcoming the challenges of transmission infrastructure in localities already vulnerable to disruption, and increased resilience. We are sceptical that the benefits of microgrids identified by some groups and individuals will be realised under the current energy governance regimes. Next, we identify and discuss the different aspirations people have for microgrids and what conditions the realisation of those aspirations might depend on. In this context we emphasise the need to focus on multiple scales, with attention to political economies and social vulnerabilities, so as to improve resilience in practice. From this basis we raise a range of skills and capabilities required of multiple actors to establish a clear idea of the future being worked towards – and then to realise it.

Realising the promises of grid-tied microgrids: A critical overview

Behind-the-meter (or distributed) energy technologies, such as solar and batteries, are increasingly popular in Australia. Around one in three Australian households has rooftop solar, together contributing 10 percent of the country's energy supply, and installations are increasing rapidly along with the average system size.[6] Home batteries, a more recently available technology, are now increasing year on year.[7] These levels of uptake have made Australia a global frontrunner and are broadly indicative of the large-scale energy transition that is underway in many parts of the world.[8] The renewable transition in Australia is largely driven by the retirement of aging coal-fired generators and the favourable economics of new solar and wind power.[9] The energy transformation from an infrastructure perspective thus involves a change from centralised, generation units located close to fossil fuel reserves, to a mix of small, renewable generators located near the point of consumption, and utility-scale renewable generators in regions with good solar and wind resources.

It would be a mistake, however, to simplify the process of energy transformation down to technologies, geographies and economics. At the local level, the transition could also be about creating futures that communities desire, underpinned by a system that is accountable to communities. For an area like the Eurobodalla Shire, the energy transition is seen by some as an opportunity to improve energy resilience and local energy sovereignty,[10] to eliminate large powerlines from their region and hence mitigate some bushfire risk, and to take action to reduce greenhouse gas emissions. A grid-tied microgrid viewed solely through an engineering lens will not bring about these desirable futures. Technologies that can help achieve these goals are the outcomes of political, economic and social factors, as are all systems that provide for the wellbeing of people.[11] The "system" that is in transition encompasses not only technologies and physical infrastructure, but the institutions that regulate, build, maintain, use and profit from it, as well as community participation and governance, local political economies, and much more. We argue that achieving the desired benefits of microgrids is contingent on processes of local capacity building and translation, which in turn requires a reorganisation of the energy system. These changes will require many actors to develop new skills and capabilities to create the conditions for transition to occur, to establish new infrastructures, and to operate and care for those infrastructures and the people who rely on them.

The potential benefits of microgrids are wide-ranging. There are technical benefits related to the ways in which "shedding" a part of the network can help keep the network stable. Beyond these technical considerations, many of the participants in this study saw microgrids as a way to alter their relationship with the energy system. Microgrids were seen as inherently good for energy security and sovereignty, building community and resilience. These benefits were seen to go hand-in-hand with the transition towards a decarbonised energy system. By localising generation and co-locating it with consumption, some participants felt that microgrids should make energy cheaper, stimulate the local economy and avoid the need for large transmission lines. We point out that these are generalised claims raised by participants, rather than tested propositions. Some of the logics behind the claimed benefits apply to decentralised energy technologies generally, so there is a need to unpack the purported benefits of microgrids and to determine what distinguishes microgrids from behind-the-meter energy technologies more generally. In this section we provide a critical overview of the desired benefits of microgrids and discuss the conditions under which microgrids might deliver these benefits.

Energy security and sovereignty

Microgrids are claimed to improve energy security and sovereignty. Energy security refers to the secure supply of energy resources. For a microgrid to provide energy security, it would have the capacity to operate during times of grid outages. This should improve day-to-day reliability and perhaps also the community's ability to recover from unpredictable, high-impact events. Energy sovereignty requires that energy policy is made primarily for the benefit of citizens as producers, consumers, account holders and stakeholders.[12] A lack of energy sovereignty indicates that energy policy is not being made in the interests of citizens, instead serving other interests. This is a contentious point, because, as some of our participants pointed out, Australia's (east coast) energy institutions are built around the assumption that a market-based approach and retail competition delivers the best outcomes for consumers based on the efficient allocation of resources. However, our research has found that householders distrust the system's profit motives and, as one participant noted, there is a perception that the system is regulated in the interests of the large, centralised, vertically integrated generator-retailer ("gentailer") incumbents.[13] There is, therefore, a disagreement between energy institutions and citizens about who energy policy is for, and who benefits from it. We hold the view that energy sovereignty refers to governance that allows for local variations in energy policy and hence takes into account local conditions and aspirations. In addition, work to build and maintain energy systems should occur locally, with the added benefit of stimulating the local economy. This requires local community participation in governance.

Reducing greenhouse gas emissions

For a grid-tied microgrid to contribute to decarbonisation, in simple terms it needs to displace more fossil fuel generation than would have been displaced in the absence of the microgrid. This requires (renewable) energy supply to be better matched to demand over the course of the day: in the middle of the day, parts of the grid have more solar generation than can be consumed, and solar power continues to grow faster than any other generation type in Australia.[14] Storage such as batteries can provide a buffer whereby excess energy is stored for later in the day when residential demand is highest, and demand management can shift consumption to times of high renewable energy generation. Into the future, microgrids could further contribute to decarbonisation by presenting new ways of distributing energy that challenge the existing paradigm and give rise to other sociotechnical changes. Proponents and participants of microgrids could contribute to long-term change by innovating and legitimising new technologies and business models to destabilise the old, acting as intermediaries to facilitate change, or as activists engaging in criticism and re-imagining the future.[15] These are all activities requiring new skills and capabilities. For example, citizens require an understanding of the affordances of energy technologies and how those affordances might work with and change their daily lives, so that they might communicate and develop a shared vision for their desired future. This requires technical knowledge and communication and collaboration skills. Groups need to be able to advocate for new rules that centre the operating modes and business models of their imagined decentralised and islandable systems, which, in turn, requires technical, governance and commercial skills. Advocacy also requires understanding the institutional and operational preferences of incumbent generators, retailers and network providers, in order to articulate what must change to realise their vision. As a community project, the successful operation of an islanded microgrid is likely to depend on people becoming more energy literate so that they can effectively participate in reducing and managing their energy demand. The implication is that with accompanying upskilling, microgrids could have a discernible role in creating momentum and reorganising rules into new configurations that favour renewable energy generation over electricity from fossil fuels.

Avoiding large transmission lines

For a microgrid to avoid the need for large transmission infrastructure implies that the community served by the microgrid has been able to attain a certain amount of independence from the grid. If not complete disconnection (in which case it would no longer be a grid-tied microgrid, but a stand-alone power system), then this independence implies a smaller powerline tethering the community to the main grid, relegating its importance in maintaining and balancing supply across the network. Such a situation could be a long-term ambition for system decentralisation: a truly bidirectional network where locally generated electricity and associated economic flows could freely move within the distribution network, incorporating local network-side infrastructure such as solar farms and large or medium-scale batteries. This scenario would only come to pass if decentralised infrastructure was a stated goal of a future energy system, which is unlikely to come to pass under the current market-based governance regime on Australia's east coast. In this scenario, the microgrid would be the primary provider of energy security, with the grid connection there to allow local energy trading and provide backup in case of localised outages. Using energy close to generation through microgrids could make a small contribution to decarbonisation by reducing losses from transporting power over long distances.

The management of a local energy system with a smaller grid connection point, and hence a greater degree of independence, brings with it greater degrees of responsibility as well as a variety of operational decisions. Electricity is an essential service, and microgrid governance regimes must have the capacity to take on responsibility and liability for supply and safety. Governance will require decisions regarding which outcomes are prioritised in operations, as microgrid managers could optimise for, say, profit, cost savings, carbon emissions or self-sufficiency.[16] Plans need to account for circumstances such as during bushfires when the connection to the main grid may be lost. During these times, governance must be equipped to successfully manage a finite, common pool of energy resources in a complex situation that could involve knowledgeable prosumers alongside less knowledgeable residents or transient tourists. Ongoing system planning would be required to decide on new investments and how to manage different trends in demand. These issues rely on not only those engaged in governance to have the appropriate skills, but also for community members to be able to participate in decisions regarding energy to a greater degree than they do presently.

Improving resilience

The potential for community microgrids to improve resilience warrants deeper discussion, as resilience is a complex topic that can tend to be oversimplified. Some interviewees working in energy policy and regulation viewed energy system resilience as similar to day-to-day reliability in energy supply, only at a greater scale with longer outages affecting more people. From this point of view, improvements to energy resilience are assessed against the cost of providing them to find the most economically efficient outcome. Other participants emphasised the goal of "bouncing back" after a major disruption and adapting to changing conditions, which requires systemic and long-term change. Some noted that community expectations of energy resilience were much higher, and encompassed more issues, than what is the currently accepted responsibility of regulators and network operators. In this section we discuss how resilience might be achieved in the context of community microgrids, severe weather and the energy transition.

Resilience in the context of energy systems usually refers to the ability of the system to recover from high impact, low probability events such as extreme weather, encompassing the immediate and long-term aftermath and focusing on not only engineering operations but on the lived experiences of people affected.[17] A resilient response could mean a return to stable conditions, either to the previous conception of "normality" or adaptation to something new.[18] Resilient energy systems should therefore have the capacity to adapt to long-term change. Because energy systems can both contribute to climate change and be subject to its impacts, long term adaptive capacity includes mitigation of climate change.[19] A microgrid designed for long-term resilience to climate change therefore should incorporate projections of the future climate, have the ability to change, and not make climate change worse.[20]

However, these are conditions for the technical design of microgrids, and microgrids are also subject to political, economic and social contexts. The goal of "bouncing back" needs to be thought of more broadly than the reinstatement of electricity service because affected people also need the capacity to adapt their own lives to rearranged conditions. For example, for the microgrid to provide the desired amount of backup, users may need to move away from being passive recipients of energy to active participants in managing demand and sticking to their "fair share," which requires understanding and capacity as well as flexibility. Likewise, the idea that resilience means a return to a previous stable state can tend to favour the status quo of existing advantage, injustice or unsustainability.[21] Instead, the governance of the microgrid must consider and constantly negotiate the relationships between community vulnerabilities and resilience.[22] A framing of resilience that is blind to unequal power relations can tend to place the onus on individuals to become more resilient and hence naturalise the neoliberal and capitalist conditions that gave rise to social inequality and climate change in the first place.[23] As some of our participants emphasised, disasters expose any weaknesses in the underlying social fabric and investments in improving social fabric are likely more effective than most technical solutions. This can be very challenging: disaster events tend to exacerbate existing household or demographic inequalities and insecurities as they intersect with sociodemographic factors, such as age, education, remoteness and reliance on social housing, that correlate with the affordability of, or interest in, energy technologies.[24] Energy vulnerabilities can also relate to network, geographical and environmental variables such as remoteness, line voltage, accessibility and cost of energy, frequency of outages and extreme weather events.[25] Social and behavioural factors such as energy practices and access to energy technologies can also shape the vulnerability of low-income households.[26] In some areas, including the NSW South Coast, transient groups such as tourists or holiday house owners bring different energy practices and needs. This can be a source of local community resentment, with visitors seen as consumers of finite resources who do not contribute.

The final point we wish to make about resilience is the need to deploy different approaches to learning and design depending on the scale of the social and technical infrastructures. In considering resilient energy systems, strategies for specific parts of the system should aim for persistence via robust design, planning and developing skills and capabilities.[27] This involves working towards set outcomes, or "single-loop learning." As the perspective broadens to larger, more complex systems, focus needs to shift to being able to cope and adapt through flexible design and collective human agency, using "double-loop learning" that questions assumptions and explores possible goals.[28] Our participants emphasised the importance of local context and experimentation in developing resilience, as well as a need for a national resilience authority with a long view and better national/state-level coordination of agencies.

Multiple levels of governance are needed in decentralised energy and could combine the small-scale benefits of local knowledge, multiple sites of innovation and cultivation of trust with the larger-scale benefits of investment and abilities to address poor or corrupt governance.[29] Likewise, governance for resilience within an organisation needs to occur at multiple scales: at the operational level, focusing on ensuring the day-to-day ability to absorb disturbances; at the tactical level, focusing on continuous improvement, adaptive risk management and opportunistic adaptive capacity; and at the strategic level, transforming the organisation towards long-term sustainability in the context of inevitable change and disruption.[30]

Working towards local grid-tied microgrids: Developing capabilities and skills

So far we have put forward conditions and complications for realising localised microgrids that deliver on a range of desired effects. Achieving these depends on different groups at multiple levels (re-)organising around new tasks and responsibilities, and novel approaches. In this section we map out further capabilities and skills required to establish and operate grid-tied microgrids. These capabilities and skills are interlinked and will apply to different actors, at different scales and at different points as the energy transition is negotiated over time.

Energy and system literacies

It is useful to begin with literacy of different aspects of the existing energy system as this provides an opportunity to describe how things currently are – a foundation for understanding what is possible within the existing regime, and then outside of it. Energy literacies begin at the most basic level: how energy is generated, transported and consumed. This can expand outwards to include literacy of broader sociotechnical systems such as rules and markets. Because microgrids put pressure on the current practices of energy provision and use, pushing them to change, proponents of microgrids need to have some understanding of what might be gained and lost through the processes of change they are advocating for. Australia's east coast National Energy Market (NEM) is an outcome of competition reforms beginning in the 1990s that shifted subnational state-owned generation, transmission and distribution infrastructures in two ways: the industry was disaggregated (distribution networks and generation were separated out), marketised and corporatised; and governance was rescaled from subnational to national.[31] As part of this process, a suite of market bodies were created that are responsible for rulemaking, implementation and enforcement, and operation of the power system and market.[32] The networks – the poles and wires – are regulated monopolies, split by state and region.

One aspect of the state-owned system that has been retained is uniform network pricing within each network business, thereby smoothing out the large variations in the cost of supply as a kind of cross-subsidy. This means that households and businesses (customers) within the same network pay the same to access the network, akin to the uniform cost of a domestic postage stamp. Energy retailers purchase energy from the wholesale market, taking on a certain amount of risk in the process, and sell it to customers, who can choose their retailer. In practice, the retailers are competing for a small proportion of engaged customers who are willing and able to "shop around" and transfer between retailers at the expense of less engaged customers.[33] People who own solar panels are paid a feed-in tariff by their retailer when they export excess solar energy to the network, which is an amount usually much lower than what they pay to purchase power – partly because solar power exports mostly coincide, making them of lower market value, and partly to preserve retailer profits.

The purported localisation benefits of microgrids depend on certain changes being made to this regime. The idea that locally generated energy should naturally be cheaper is one with likely social regressive consequences. One reason that locally generated energy might be cheaper is because it does not have to travel very far, which should lead to lower network costs. For customers to pay lower costs, however, networks must be allowed to break with uniform pricing, thereby exposing other customers, particularly those in remote parts of the grid, to higher costs. Grid-tied microgrids are part of the main grid most of the time, meaning the microgrid users would no longer be sharing the collective cost of the network. The second way that local energy might become cheaper is if economic flows were as free as the electrons, and a unit of energy cost the same regardless of whether it was purchased from a retailer or from a householder. This is a reordering of the roles of "retailer" and "customer."

Localisation could challenge the profit motive of the market, a source of simultaneous order and distrust. However, there would still be a connection to the main grid and hence the NEM. If the microgrid operated by its own set of rules while islanded, there would likely be no material change to the microgrid users' energy costs because islanding happens so seldom. Alternatively, the microgrid could be an "embedded network" whereby users of the microgrid trade energy into and from a local pool that is a single customer in respect of the NEM. The main issue with an embedded network is that all customers within the microgrid are a part of it by default (unless they are motivated and have the capability to opt out and install their own meter). Embedded networks require special regulation because the users are no longer covered by the consumer protection systems of the retail market. The least complicated option is for the microgrid to continue operating by the rules of the NEM while islanded, which means pretending the market is functioning normally for the duration of the outage. This, however, rules out localisation benefits that might have been the reason people wanted the microgrid in the first place. The point is that following through with the simple proposition that microgrids should make energy cheaper creates cascading problems for proponents and participants alike to grapple with, including people who do not understand how the systems of accountability have been set up.

Design and operation

As we described earlier, microgrids are designed to optimise for specific outcomes, such as maximising the use of locally produced energy or maintaining a certain duration of back-up power.[34] The possibility of multiple optimisation goals is just one among many sociotechnical factors that make technologies like microgrids inherently complex. Digitalisation in the form of software algorithms and automation is proposed as a solution to multiple objectives, but digitalisation in practice can hide complexity, leading to even less clarity, missed communications and a decline in trust.[35] In other words, putting an algorithm in charge of decision-making can create new complexity that must then be managed through continual social interventions such as adjustment, information-sharing and evaluation.[36] The range of necessary skills to deliver a successful microgrid proliferates.

To add yet more complexity, a grid-tied microgrid works in multiple modes. Most of the time it would be expected to work in grid-tied mode, requiring the lowest level of community participation in managing household energy use. In islanded mode, users of the microgrid might need to be more judicious with their use of electricity in order to maintain a finite supply for as long as needed. But if the power has not been interrupted – because there is a microgrid – how will consumers know it is time to practice a different mode of consumption? Furthermore, there may be an additional mode when the community is preparing for an approaching disruption. During this time, microgrid users might need to conserve energy. Automation could offer a solution, but microgrid designers still need to appreciate the additional social capabilities and processes required, and operators will need to make it work.

Community deliberation and coordination

At the heart of this topic is the need for well-developed skills for deliberating, collaborating and conversing with communities, as well as within communities. As we have argued, in order to realise the promises of grid-tied microgrids, energy users – whether householders or businesses – must be the primary beneficiaries, in terms of both observable outcomes and perception. Microgrid systems need to be tailored to local conditions rather than inheriting the default objectives of the broader grid. Local communities must have a role in making decisions on the types of technologies being used and the extent of participation that is required or expected, as well as in defining, advocating for and building momentum towards their desired energy futures. Inherent to all of these processes are skills and capabilities in democratic community engagement and coordination.

Local-scale, face-to-face interactions are recognised as essential for cultivating trust and a shared motivation to maintain relationships.[37] One participant emphasised the role of community champions in fomenting trust, by acting as intermediaries between people and bureaucracies. Local context also needs to be incorporated into decision-making, complete with understanding of the vulnerabilities present in the community and in a manner that facilitates constructive negotiation between different interests, including groups who are not directly represented.[38] This must include non-locals as well as future generations.[39] Fair representation of interests is of particular importance for local energy projects where, as some of our participants noted, representation tends to skew to wealth and property ownership, higher education, and general loudness. The need for ongoing deliberation over interests and priorities as well as translating these to operational outcomes also runs counter to the "set and forget" operational models that are typical of energy projects.

Aside from community capacity, we also argue that institutional actors and community energy groups require more sophisticated skills in community deliberation and engagement. If these actors are unable to effectively coordinate open dialogues and negotiations with people and effect their participation in energy systems, poor outcomes will emerge: the grid-tied microgrids could be subsumed into the existing logics and objectives of the wider grid, delivering little perceptible difference to the community. Alternatively, they could give responsibility to individuals and regressively funnel benefits to the most engaged participants. Ideally, everyday householders and other energy users would, through accessible material engagement with the microgrid, see the relevance of it to their everyday lives and see their interests represented in its governance. Engagement needs to include mapping of the community, its power asymmetries and accountability gaps, with competencies in sensitive facilitation and community development. We note that many of these skills already exist within some local government organisations; however, we found from our interviews that local governments did not have the resources to get involved in the issue of energy supply beyond emergency coordination and general advocacy.

Governance and accountability

Throughout this discussion, we have raised the importance of governance in realising the promises of grid-tied microgrids. By governance, we mean the processes, systems and actors involved in making and enforcing rules that address collective action problems, deal with market failures and ensure the provision of public goods.[40] Governance institutions operate at different levels, ranging from state or national rulemaking, to the rules governing the settings of operational systems and rules affecting day-to-day operations.[41] Governance also applies within organisations ranging from strategic to operational decisions. Localised microgrids therefore need to be accommodated in rules and enforcement at the national scale, as well as at the scale of the microgrid itself. Systems of governance that extend beyond state actors include nodes of decision-making that are not subject to democratic mechanisms.[42] Because of this, particular attention must be paid to the types of checks and balances used to maintain adequate accountability.[43] This applies to Australia's energy system since the reforms of the 1990s and increasingly so as decentralisation occurs. The decentralisation of governance is ideally participatory, and devolved to the lowest appropriate level, but also where regulators are able to apply formal sanctions and where groups have sufficient capacities, financing and powers to govern.[44] The national bodies that oversee rulemaking, rule implementation and enforcement, and system operation therefore need to develop ways of governing a participatory, decentralised system. For such techno-economically oriented institutions, developing these types of social capacities would require organisational reflexivity and transformation.

Community groups engaged in participatory governance require the ability to interact with, mobilise and activate community assets towards a common purpose.[45] One of our participants made the point that serious training in these "commoning" skills would be required for South Coast community energy groups. Another argued that practising face-to-face discussion between community members was an essential and irreplaceable part of cultivating trust and accountability. Community groups that take on responsibilities in delivering essential services like electricity also need to be sustainable. They must avoid placing too much reliance on individuals, as volunteers are prone to burnout, a situation that becomes more difficult during times of economic decline.[46] Groups must have the skills to negotiate community interests that may be conflicting and must work transparently so that larger scales of governance are able to monitor for potential problems such as local tyrants, non-contributors, negative discrimination and strategic behaviour.[47]

A key question emerges: to what extent can or should communities be involved in governing grid-tied microgrids? Our participants largely agreed that the community should be involved in decision-making and design, but differed in the degree to which they see this involvement being productive. Unsurprisingly, the institutional actors who regulate and operate the energy system appeared to envisage the least active role for community members, only seeing them through an economic lens as "consumers" who participate via a retail offering. Others recognised the need for community engagement in energy system planning, but preferred that it occur under a "residual realist" framework whereby the engagement format is fixed, designed to be scalable, and occurs after the technical challenges have been solved.[48] They framed these as pragmatic processes within the constraints of conventional engineering project delivery. All participants – including people with experience in local community organising – agreed there are limits to community ownership and governance because of limited capacity to manage the complexity, legal liability and safety of energy infrastructure. However, some recognised that community groups might be in a position to take on extra risks and tolerate failure, in contrast to other actors, such as local governments, who must apply more pragmatism and risk-aversion to investments and decisions.

Conclusion: Creating collective energy futures

Throughout this chapter we have raised a daunting set of skills and capabilities required for localised grid-tied microgrids to realise their potential benefits of energy security and sovereignty, reducing greenhouse gas emissions, avoiding long transmission lines and improving community resilience. We have found the need for literacy in energy as well as the broader energy system, and technical skills in the design and operation of software, hardware and networks. On the more social side, skills in engaging community members in the material realities of the microgrid, ensuring the broad representation of interest and participation in governance are needed. Our interviews have revealed shortfalls in governance frameworks that ensure communities can build microgrids in line with the values of a sustainable energy system, in particular addressing questions of equity and justice. Without this it will be difficult to answer or resolve practical questions such as, in what locations would a grid-tied microgrid improve energy resilience? Further, what other capacities and activities would be required to improve community energy resilience? Not to mention resolving whether alternatives to microgrids may prove to be more suitable to improve resilience in environmentally sensitive locations.

Realistically, however, the energy transition is a long-term, complex unfolding process that contains great uncertainty about the future. Microgrid proposals could (with appropriate support) become sites of local experimentation, learning, visioning and practising commoning skills, all of which are necessary processes in exploring possibilities for the future and stimulating change. In Australia, the decentralisation of energy has been led by individual householders and community groups who are frustrated with the slow pace of the energy transition. If we zoom out to fully understand householder interest and excitement in technologies such as neighbourhood batteries and microgrids, as well as the vulnerabilities communities face in a changing climate, we can more fully begin to grapple with the nature of the governance challenge before us. The motivations, interests and capacities of large-scale market stakeholders (for whom the current system is designed) vary greatly from the householders we have spoken to. Likewise, the generalised logics and objectives of the large-scale market grate against the unique circumstances and desires of individual communities. Extreme events such as the Black Summer will continue to occur, and communities will keep investigating novel alternatives. New models of governance, ownership and operation will need to develop to bridge the divide and bring these two scales of energy system into something resembling a cohesive, equitable and sustainable whole.

Acknowledgements

Thanks are due to the editors and Clare Silcock for feedback gratefully received. The work presented in this chapter was conducted as part of the Southcoast μ-grid Reliability Feasibility (SμRF) project, funded through the Regional and Remote Communities Reliability Fund of the Australian Department of Climate Change, Energy, the Environment and Water.

Notes

1 Commonwealth of Australia, *Royal Commission into National Natural Disaster Arrangements Report.*
2 Essential Energy, "Electricity Network Safety Management System Performance & Bushfire Preparedness Report 2019–20."
3 NSW Government, "2022 Flood Inquiry - Volume Two: Full Report."

4 Australian Photovoltaic Institute, "PV Postcode Data."
5 Eurobodalla Shire Council, "Submission to NSW Government Regional Housing Taskforce."
6 Clean Energy Council, "Clean Energy Australia Report 2023"; DCCEEW, "Australian Energy Statistics."
7 Clean Energy Council, "Clean Energy Australia Report 2023."
8 Blakers et al., "Pathway to 100% Renewable Electricity."
9 Australian Energy Regulator, *State of the Energy Market 2020*.
10 As we will expand upon later, energy sovereignty refers to the positioning of local residents and businesses as the primary beneficiaries of energy governance.
11 Delina, Ocon, and Esparcia, "What Makes Energy Systems in Climate-Vulnerable Islands Resilient?"
12 In using the term "citizens" we do not mean to exclude people without state-issued citizenship. Underwood, Hill, and Lamichhane, "Earthquakes, Blockades and Energy Crises"; Szulecki, "Conceptualizing Energy Democracy."
13 Ransan-Cooper et al., "Applying Responsible Algorithm Design."
14 Clean Energy Council, "Clean Energy Australia Report 2023."
15 Wilkinson et al., "Is Peer-to-Peer Electricity Trading Empowering Users?"
16 Ransan-Cooper et al., "Applying Responsible Algorithm Design to Neighbourhood-Scale Batteries in Australia."
17 Mishra et al., "A Review on Resilience Studies in Active Distribution Systems."
18 Nik, Perera, and Chen, "Towards Climate Resilient Urban Energy Systems."
19 Kelly-Pitou et al., "Microgrids and Resilience."
20 Kelly-Pitou et al., "Microgrids and Resilience."; Nik, Perera, and Chen, "Towards Climate Resilient Urban Energy Systems."
21 Harris, Chu, and Ziervogel, "Negotiated Resilience."
22 Delina, Ocon, and Esparcia, "What Makes Energy Systems in Climate-Vulnerable Islands Resilient?"
23 DeVerteuil and Golubchikov, "Can Resilience Be Redeemed?"; MacKinnon and Derickson, "From Resilience to Resourcefulness."
24 Chen et al., "Extreme Events, Energy Security and Equality."; Nicholls, Strengers, and Tirado, "Smart Home Control"; Sovacool et al., "Vulnerability and Resistance in the United Kingdom's Smart Meter Transition."
25 Day, Walker, and Simcock, "Conceptualising Energy Use and Energy Poverty"; Liévanos and Horne, "Unequal Resilience"; Winzer, "Conceptualizing Energy Security."
26 Chen et al., "Extreme Events, Energy Security and Equality."
27 van der Merwe, Biggs, and Preiser, "A Framework for Conceptualizing and Assessing the Resilience of Essential Services."
28 van der Merwe, Biggs, and Preiser, "A Framework for Conceptualizing and Assessing the Resilience of Essential Services."; Hamborg et al., "Rethinking Resilience."
29 Brisbois, "Decentralised Energy, Decentralised Accountability?"; Ostrom, "Polycentric Systems."
30 van der Merwe, Biggs, and Preiser, "A Framework for Conceptualizing and Assessing the Resilience of Essential Services."
31 Chandrashekeran, "Re-Stating Power."
32 Respectively, these are the Australian Energy Market Commission, the Australian Energy Regulator, and the Australian Energy Market Operator.
33 MacGill and Smith, "Consumers or Prosumers, Customers or Competitors?"
34 Ransan-Cooper et al., "Applying Responsible Algorithm Design."
35 Hansen et al., "Smart Technology Needs Smarter Management."
36 Hansen et al., "Smart Technology Needs Smarter Management."
37 Ansell and Gash, "Collaborative Governance in Theory and Practice."

38 Harris, Chu, and Ziervogel, "Negotiated Resilience."
39 Bradshaw, "Questioning the Credibility and Capacity of Community-Based Resource Management."
40 Sovacool and Florini, "Examining the Complications of Global Energy Governance"; Young, "Land Use, Environmental Change, and Sustainable Development."
41 Ostrom, *Understanding Institutional Diversity*.
42 Brisbois, "Decentralised Energy, Decentralised Accountability?"
43 Brisbois, "Decentralised Energy, Decentralised Accountability?"
44 Brisbois, "Decentralised Energy, Decentralised Accountability?"
45 Brisbois, "Decentralised Energy, Decentralised Accountability?"
46 Bradshaw, "Questioning the Credibility and Capacity of Community-Based Resource Management."
47 Brisbois, "Decentralised Energy, Decentralised Accountability?"
48 Chilvers, Pallett, and Hargreaves, "Ecologies of Participation in Socio-Technical Change."

Bibliography

Ansell, Chris and Alison Gash. "Collaborative Governance in Theory and Practice." *Journal of Public Administration Research and Theory* 18, no. 4 (2008): 543–71.

Australian Energy Regulator. *State of the Energy Market 2020*. Melbourne: Australian Competition and Consumer Commission, 2020.

Australian Photovoltaic Institute. "PV Postcode Data. 2023." https://pv-map.apvi.org.au

Blakers, Andrew, Matthew Stocks, Bin Lu, Cheng Cheng and Ryan Stocks. "Pathway to 100% Renewable Electricity." *IEEE Journal of Photovoltaics* 9, no. 6 (2019): 1828–33.

Bradshaw, Ben. "Questioning the Credibility and Capacity of Community-Based Resource Management." *Canadian Geographer* 47, no. 2 (2003): 137–50.

Brisbois, Marie Claire. "Decentralised Energy, Decentralised Accountability? Lessons on How to Govern Decentralised Electricity Transitions from Multi-Level Natural Resource Governance." *Global Transitions* 2 (2020): 16–25.

Chandrashekeran, Sangeetha. "Re-Stating Power: How States Make, Tame, and Shape Markets." *Geographical Research* 60, no. 1 (2021): 138–51.

Chen, Chien-fei, et al. "Extreme Events, Energy Security and Equality through Micro- and Macro-Levels: Concepts, Challenges and Methods." *Energy Research and Social Science* 85 (2022): 102401.

Chilvers, Jason, Helen Pallett and Tom Hargreaves. "Ecologies of Participation in Socio-Technical Change: The Case of Energy System Transitions." *Energy Research and Social Science* 42 (2018): 199–210.

Clean Energy Council. "Clean Energy Australia Report 2023." Melbourne: Clean Energy Council, 2023. https://assets.cleanenergycouncil.org.au/documents/Clean-Energy-Australia-Report-2023.pdf

Commonwealth of Australia. *Royal Commission into National Natural Disaster Arrangements Report*, Canberra, 2020.

Day, Rosie, Gordon Walker and Neil Simcock. "Conceptualising Energy Use and Energy Poverty Using a Capabilities Framework.' *Energy Policy* 93 (2016): 255–64.

Delina, Laurence L., Joey Ocon and Eugene Esparcia. "What Makes Energy Systems in Climate-Vulnerable Islands Resilient? Insights from the Philippines and Thailand." *Energy Research and Social Science* 69 (2020): 101703.

Department of Climate Change, Energy, the Environment and Water (DCCEEW). "Australian Energy Statistics, Table O Electricity Generation by Fuel Type 2022–23 and 2023." 23 April, 2024. https://www.energy.gov.au/publications/australian-energy-statistics-table-o-electricity-generation-fuel-type-2022-23-and-2023

DeVerteuil, Geoff and Oleg Golubchikov. "Can Resilience Be Redeemed?" *City* 20, no. 1 (2016): 143–51.

Essential Energy. "Electricity Network Safety Management System Performance & Bushfire Preparedness Report 2019–20," 2020. https://www.essentialenergy.com.au/-/media/Project/EssentialEnergy/Website/Files/Our-Network/ENSMS-Annual-Report-2019-20.pdf?la=en&hash=9A80D3623163E1527F6901011699A73BE6D01064

Eurobodalla Shire Council. "Submission to NSW Government Regional Housing Taskforce," 2021. https://www.esc.nsw.gov.au/__data/assets/pdf_file/0019/225910/Submission-to-NSW-Govt-Regional-Housing-Taskforce.pdf

Hamborg, Steffen, Jasper N. Meya, Klaus Eisenack and Thorsten Raabe. "Rethinking Resilience: A Cross-Epistemic Resilience Framework for Interdisciplinary Energy Research." *Energy Research and Social Science* 59 (2020): 101285.

Hansen, Paula, Gregory M. Morrison, Atiq Zaman and Xin Liu. "Smart Technology Needs Smarter Management: Disentangling the Dynamics of Digitalism in the Governance of Shared Solar Energy in Australia." *Energy Research and Social Science* 60 (2020): 101322.

Harris, Leila M., Eric K. Chu and Gina Ziervogel. "Negotiated Resilience." *Resilience* 6, no. 3 (2018): 196–214.

Kelly-Pitou, Katrina M., et. al. "Microgrids and Resilience: Using a Systems Approach to Achieve Climate Adaptation and Mitigation Goals." *The Electricity Journal* 30, no. 10 (2017): 23–31.

Liévanos, Raoul S. and Christine Horne. "Unequal Resilience: The Duration of Electricity Outages." *Energy Policy* 108 (2017): 201–11.

MacGill, Iain and Robert Smith. "Consumers or Prosumers, Customers or Competitors? Some Australian Perspectives on Possible Energy Users of the Future." *Economics of Energy and Environmental Policy* 6, no. 1 (2017): 51–70.

MacKinnon, Danny and Kate Driscoll Derickson. "From Resilience to Resourcefulness: A Critique of Resilience Policy and Activism." *Progress in Human Geography* 37, no. 2 (2013): 253–70.

Mishra, Dillip Kumar, et al. "A Review on Resilience Studies in Active Distribution Systems." *Renewable and Sustainable Energy Reviews* 135 (2021): 110201.

Nicholls, Larissa, Yolande Strengers and Sergio Tirado. *Smart Home Control: Exploring the Potential for Enabling Technologies in Vulnerable, Disengaged and Regular Households*. Centre for Urban Research, RMIT University, 2017.

Nik, V. M., A. T. D. Perera and D. Chen. "Towards Climate Resilient Urban Energy Systems: A Review." *National Science Review* 8, no. 3 (2021): nwaa134.

NSW Government. "2022 Flood Inquiry - Volume Two: Full Report," 29 July 2022.

Ostrom, Elinor. "Polycentric Systems for Coping with Collective Action and Global Environmental Change." *Global Environmental Change*, 20th Anniversary Special Issue, 20, no. 4 (2010): 550–57.

Ostrom, Elinor. *Understanding Institutional Diversity*. Princeton: Princeton University Press, 2005.

Ransan-Cooper, Hedda, Björn C. P. Sturmberg, Marnie E. Shaw and Lachlan Blackhall. "Applying Responsible Algorithm Design to Neighbourhood-Scale Batteries in Australia." *Nature Energy* 6 (2021): 815–23.

Sovacool, Benjamin K., and Ann Florini. "Examining the Complications of Global Energy Governance." *Journal of Energy and Natural Resources Law* 30, no. 3 (2012): 235–63.

Sovacool, Benjamin K., Paula Kivimaa, Sabine Hielscher and Kirsten Jenkins. "Vulnerability and Resistance in the United Kingdom's Smart Meter Transition." *Energy Policy* 109 (2017): 767–81.

Szulecki, Kacper. "Conceptualizing Energy Democracy." *Environmental Politics* 27, no. 1 (2018): 21–41.

Underwood, G., D. Hill and S. Lamichhane. "Earthquakes, Blockades and Energy Crises: A Conceptual Framework for Energy Systems Resilience Applied to Nepal." *Energy Research and Social Science* 69 (2020): 101609.

van der Merwe, Susara E., Reinette Biggs and Rika Preiser. "A Framework for Conceptualizing and Assessing the Resilience of Essential Services Produced by Socio-Technical Systems." *Ecology and Society* 23, no. 2 (2018). http://www.jstor.org/stable/26799110

Wilkinson, Sam, et al. "Is Peer-to-Peer Electricity Trading Empowering Users? Evidence on Motivations and Roles in a Prosumer Business Model Trial in Australia." *Energy Research and Social Science* 66 (2020): 101500.

Winzer, Christian. "Conceptualizing Energy Security." *Energy Policy* 46 (2012): 36–48.

Young, Oran. "Land Use, Environmental Change, and Sustainable Development: The Role of Institutional Diagnostics." *International Journal of the Commons* 5, no. 1 (2011): 66–85.

3

NAVIGATING SKILLS IN UNCERTAIN ENERGY FUTURES

Education, Skills and Training for the Australian Air-Conditioning Workforce

Elyse Stanes, Chantel Carr, Pauline McGuirk, Matthew Daly and Daniel Daly

Introduction

William, Edward, Noel and Neil meet online each Wednesday evening, after dinner and time with their families, to discuss changes in the Certificate III in Air-Conditioning and Refrigeration (hereafter Certificate III), a trade qualification in Australia's Vocational Education and Training (VET) sector.[1] All experienced teachers, they have held late night conversations for the past six months to discuss the teaching frameworks and assessment materials that they will collectively propose to their workplaces for the next iteration of the Certificate III. Their aim, according to Noel, is to collaborate as curriculum managers "so that we could have more force to [do] what we want and how to do it." The Certificate III in Air-Conditioning and Refrigeration is Australia's nation-wide and formal training route for "fridgies" or workers in the Heating, Ventilation, Air-Conditioning & Refrigeration (HVAC&R) sector. It is the only formal training that the majority of the HVAC&R workforce will receive over their careers.[2]

HVAC&R infrastructure underpins thermal comfort, globalised food supply chains, healthcare, agriculture, and many other industrial applications. While largely invisible, HVAC&R infrastructure is ubiquitous and essential to modern life. Demand is also growing. As climate change accelerates, the International Energy Agency (IEA) estimate that global demand for space cooling will more than triple by 2050, labelling it "one of the most critical energy issues of our time."[3] Already consuming ten percent of global electricity supply[4] and relying on planet-warming gases that end up in the atmosphere when not handled and disposed of properly, HVAC&R infrastructure is at once both vital and environmentally challenging. It requires a skilled and

DOI: 10.4324/9781003512240-5

robust workforce to manage and maintain it, and to deliver crucial performance improvements in energy efficiency.

Teachers like William, Edward, Noel and Neil have had long careers on the frontline of HVAC&R trades training, including navigating changes to education frameworks as well as industrial and technological change. Given Australia's extensive experience with extreme heat and increasing frequency of heatwaves, their perspectives offer globally relevant insights for addressing immediate challenges and informing future climate adaptation strategies. As the primary and (across careers) likely only formal educators of HVAC&R skills,[5] VET teachers are also critical to the development of the skilled workforce that will be required to address future energy transitions in the sectors outlined above.

The vocational context in which HVAC&R teachers work faces challenges in responding to the multi-dimensional nature of the climate crisis. This is in part due to the traditional pillars that the VET sector addresses – employability, skills shortages and work practices – being held apart from the broader sustainability challenges that also have impacts on trade industries.[6] A compounding tension for the HVAC&R sector is pipeline issues for bringing new entrants into the workforce. The former Australian Industry Skills Committee estimated that approximately 30,000 people were employed as "fridgies" in 2021.[7] Projections at the time indicated that an additional 4,500 fridgies would be needed by 2025, mobilising concern around skills shortages in a largely invisible sector. This growth in demand has implications for trades training, raising questions about exactly who will train the next generation of apprentices. Demand for trades is further intensified by Australia's decarbonisation agenda, with HVAC&R workers crucial to improving energy efficiency and reducing emissions across the built environment. While programmes for retrofitting the built environment, for example, are rapidly escalating, they will need to accelerate even further to meet net zero commitments.

This chapter extends recent scholarship on the workforce implications of transitioning to low-carbon futures by foregrounding the work of VET teachers responsible for formal training and skills development. Our specific focus is the experiences of HVAC&R VET teachers tasked with the professional development of the "fridgie" workforce. We adopt geographer Sarah Mills's focus on *training* as a particular type of learning, described as "skills-based, staged, repeated, refined, observed, assessed."[8] While the training required to become a "fridgie" essentially takes place across formal and informal settings throughout a career, it is the sites of formal vocational training that anchor where and indeed what trade skills are acquired at the outset.

Extending on Mills, we also consider the work of "trainers." We contend that, within formal sites of training, VET teachers become important "middle actors" who mediate across learning, training and work, curriculum interpretation and negotiate access to sectors and workplaces.[9] In HVAC&R, VET teachers are uniquely and influentially positioned between educational governance, industry and the workforce receiving education and training (including incoming apprentices). They also serve as important middle actors in navigating the needs and demands of uncertain futures, whether that be in response to changing educational or industry demands, to evolving regulations or standards in built environment energy performance in response to climate change, or adapting to environmental crisis. We argue that their work has been under-appreciated and undervalued, particularly their resourcefulness in meeting changing workforce and skills demands associated with future energy transitions. As societies continue to navigate uncertain futures, this kind of resourcefulness, we argue, will be much needed.

This chapter takes a relational approach towards "trainers" and sites where "training" happens. We look to middle actors as a vital resource for understanding issues of skills gaps, industrial and technological change, challenges for workforce attraction and retention, and environmental pressures. How might the *resourcefulness* of VET teachers assist with bridging gaps or coping with lags while such transitions take place? And equally, we look to formal sites of vocational training as an anchor for practical learning in uncertain futures. How might teachers work with vocational sites of education and training to support future energy transitions? Both questions have important implications for how we frame and value the knowledge, skills and influence of those engaged in training the HVAC&R workforce of the future.

The empirical work that underpins this chapter arose from a two-year project commissioned by the former Australian government Department of Industry, Science, Energy and Resources (DISER). An overarching aim of the broader project was to look at repair and maintenance practices as a route to improving energy efficiency in the built environment.[10] Our interdisciplinary project team engaged with a range of people working in the HVAC&R sector about their experiences on the ground, including how their everyday work practices impact on energy efficiency.[11] Looking across the sector and beyond, we identified skills and training as a current and future sticking point for energy efficiency gains in the built environment. While our research did not explicitly set out to speak with VET teachers about their careers, nor how they navigated different educational challenges, interviews with those working in the HVAC&R education sector revealed their deep commitment to skills development, often in the face of a system more oriented towards commercial priorities.

The following section positions this chapter within relevant literature on the geographies of work and skill in the face of uncertain futures. From there, we provide an overview of the specific Australian apprenticeship context for the HVAC&R sector and the "fridgie" workforce, as well as the context in which HVAC&R VET teachers work. The chapter then shifts to the empirical material. It first details how HVAC&R teachers use their own knowledge and skills within an increasingly rigid VET context, before following how teachers are working to transform formal sites of education via their own communities of practice. These signs of resourcefulness, we argue, offer instructive opportunities to improve education pathways for a workforce on the frontline of future energy transitions in the built environment.

Navigating work and skill in uncertain futures

The future of work has become a trending topic in recent discussions on energy transformations. Some predict jobs in carbon-intensive sectors will be displaced by a shift to new "green" economies with little input from workers themselves.[12] Others foresee technological changes creating new jobs in more highly skilled and safer work contexts.[13] Both scenarios suggest significant shifts in skills demand. Concerns over meeting these demands have drawn interest from education, government and regulatory bodies, each with attempts to standardise and formalise training with future needs in mind.[14] Across these discussions, the question of how energy transitions will re-shape future employment, labour markets and skills demand has been a central concern.

In geography – where this research is largely positioned – academic responses to the future of work and skills have challenged the tendency to overlook existing workforce capabilities in favour of new expertise, contending that current debates ought to "look *within*," as well as beyond, for clues on how we might adjust to "contingency and disruption."[15] Examining who will perform the work of climate crisis raises questions about what "skills might be needed, adapted or resurrected, in order to meet the pressing challenges that lie ahead."[16] As Chantel Carr has argued in previous work, and in the Introduction to this volume (with Jesse Adams Stein), rather than devaluing the skills of existing carbon-intensive trades workers – like those in the HVAC&R sector – they could be understood as valuable assets "for other kinds of material repurposing, repairing cultures."[17] Such skills, Carr argues, are essential for navigating future uncertainties.[18]

The acquisition of skill itself requires that there are opportunities for practical and hands-on learning.[19] Practical learning involves the acquisition of "know-how," or the "capacity to engage reflexively in personally and socially valued embodied performances in accordance with the demands of specific situations."[20] It is widely recognised that when learning or developing skills, the learning process (or training) is not a singular, unified experience, but occurs across formal and informal settings, and through "multi-scalar convergence of actors, interests, incentives, relationships, institutions and materials."[21] Participation in such systems underpins what anthropologist Jean Lave and educational theorist Etienne Wenger coined "communities of practice."[22] While contingent and fragile, the coming together of people, sites and practices also imply that becoming skilled is a processual, "never-ending experimental engagement."[23]

Thinking about how new and existing skills might be held together also means looking at the micro-spaces of existing training cultures to understand how, and with what knowledge, new capacities and capabilities might develop in response to uncertain futures. Most studies of communities of practice either look across formal and informal settings, or focus solely on the role of informal networks.[24] Sites of formal training – like vocational settings – are often neglected. Rather than a focus on skills, there is a tendency in more formal training settings to prioritise curricular development, qualification frameworks and institutional reform ahead of practical learning or "vocational pedagogy."[25] Our contribution here is to examine skill, learning and training through formal sites, via the perspectives of the actors who are actively involved in the transfer of knowledge in VET institutions.

In VET institutions, teachers are responsible for translating national competency frameworks into practical learning outcomes. Effectively, they are the middle actors between top-down educational governance frameworks, and the knowledge imparted to apprentices.[26] While VET teachers work within formal institutions, they also act as intermediaries across the learning-training-workplace nexus, applying their own ideas in shaping decisions about what is taught and importantly, *how* it is taught. We recognise that this is not always straightforward. Teachers themselves evidently need time, resources, social and industry networks, and a facilitative regulatory environment, to organise enriching practical classes.[27] As we detail in the following section, opportunities in the Australian VET system are conditioned by the broader economic and political environment. Examining how teachers navigate this context – alongside a changing industry landscape – provides a unique entry point for how their roles connect training provision with future needs and demands of the HVAC&R sector. We turn now to the HVAC&R training regime in Australia, where vocational education combines formal qualifications (Certificate III) with apprenticeships.

Training the fridgie workforce: A background to air-conditioning and refrigeration apprenticeships and the VET sector

In the Australian apprenticeship system, HVAC&R apprentices undertaking the Certificate III are typically simultaneously contracted to an employer for four years. The intention is that during this time, skills and knowledge are passed on from employer to apprentice.[28] But the apprentice experience can vary a great deal depending on the size of the company employing them, the type of work the firm carries out, or indeed, the level to which an apprentice is taken "under the wing" of an employer or supervisor. Across our interviews with the HVAC&R workforce, it was not uncommon to hear of apprentices who spent their entire time in training working on "low-skill" tasks: that is, employers vary greatly in terms of how they train their apprentices. In an air-conditioning and refrigeration apprenticeship, these typically involve repetitive tasks such as the cleaning and replacement of filters or installation of split system air-conditioners.

The huge variation in the apprenticeship experience in HVAC&R means that VET training programmes are vital in developing a consistent training standard for apprentices.[29] VET institutions aim to provide necessary exposure to the broad range of technologies and work processes beyond those of a single employer. Across a four-year HVAC&R apprenticeship programme, it is intended that apprentices will develop expertise in various aspects of this diverse and highly skilled trade including domestic and commercial applications, installing and maintaining mining or agricultural machinery, and handling the range of different refrigerants in service. Importantly, one key purpose of trade education in the Australian VET system is to provide a connection between industry needs and trade skill demand. As such, these formal sites of training are central in the intermediation between educational frameworks for skills development and competence, and changing workforce demands as expectations (and regulation) of energy performance tighten across many applications, including the built environment, agriculture and food supply chain.

Historically, the Australian VET sector was built on dual aims: empowering working-class students through education while meeting industry skills needs.[30] However, decades of neoliberal reforms have transformed the sector through multiple, reinforcing pressures: market deregulation and privatisation, chronic institutional underfunding, the closure of programmes and loss of experienced staff. Simultaneously, Australia's deindustrialisation trajectory has reduced overall apprenticeship numbers, creating a downward spiral where declining enrolments trigger further funding cuts.[31] These systemic changes have transformed teaching conditions and institutional capacity.

The impact of these reforms is visible in resourcing differences across Australia. Metropolitan VET institutions are often equipped with state-of-the-art systems fit for purpose for the contemporary workforce. Like many of the metropolitan-based teachers we interviewed, William taught his apprentices on a wide range of sophisticated and technologically advanced equipment:

> They've got a lot of stuff. There's a supermarket, like a little mini-supermarket. And a blast freezer. We've got all sorts of stuff, and some of the apprentices, they just meander around, and it's like, guys, when I did my apprenticeship a long time ago, we just had this one little unit and 15 guys around it.

This sat in contrast to the regional institution where Edward worked. Here, smaller class sizes and budget constraints meant that students might not get the full range of experience across their apprenticeship:

> We've got one cool room, one freezer room. I don't have a single commercial cabinet, which annoys the absolute blazes out of me. I've been beating the daylights out of my bosses for, I don't know how long, to get some commercial equipment. But it's money they can't afford to lay out for 50 students, so we struggle a little bit.

These institutional inadequacies mask an important dynamic: while the VET sector struggles to provide basic resources, the role HVAC&R teachers play as middle actors bridges growing gaps between system constraints and emerging educational and industry needs. Their efforts to maintain quality training despite these constraints often obscures the extent of institutional decline, even as they shoulder increasing responsibility for delivering the skills needed for future energy transitions. As middle actors, HVAC&R teachers are mediating the delivery of education and training in resource-constrained environments, while at the same continuously realigning with evolving skills demand.

The career history and context of HVAC&R teachers is also important. All teachers we spoke with had undertaken the same form of apprenticeship outlined above. The vast majority of teachers returned to vocational education after a period of time "on the tools," where they honed skill in particular areas of this diverse sector. Some were still taking on work on the side, in additional to their teaching roles. In our interviews, teachers often confessed that they did not plan a career in vocational education. Rather, teaching was something that appeared at the right time in their personal and professional lives. Most participants had also started teaching in the 1980s, before market-driven deregulation and privatisation reshaped the VET sector. The cohort of teachers that we spoke with were, by their own admissions, in the later stages of their career. As Tim describes, the HVAC&R teaching workforce is an aging sector: "The age of the average teacher in New South Wales, he's close to 60 and they're saying, well, are we going to be able to continue?"

All believed that younger teachers were needed to support the skills challenges the trade faces into the future. According to Tim, in 2021 there were 122 HVAC&R educators working across 26 VET institutions in Australia. The average age of a HVAC&R teacher is estimated to be upwards of 55 years of age. Several teachers were still teaching into their seventies. Recruiting new teachers to the profession is difficult, in part because of the severe defunding of the sector, outlined above. While an aging cohort of HVAC&R teachers is problematic for an industry at the front line of increasingly rapid environmental, technological and economic change, their maturity and experience also make them vital middle actors. This positions them uniquely to shape skills development and improve educational outcomes, and navigate the uncertainties of future energy transitions.

While HVAC&R teachers have been at the forefront of both historical and contemporary contexts of aligning training and education with workforce skills demands, almost no work has sought to understand the centrality of their role as intermediaries or valued their insights on what will need to be addressed in the sector to meet emergent demand, especially in the context of climate change. In the next section, we begin to unpack this. What emerges is a complex picture that reveals the risks of an overburden of compliance, the tensions between teachers' sense of good pedagogical practice and the impacts of funding constraints on quality of teaching, and the impact these have on the development of future skills needed in the HVAC&R sector.

Skills and training pressures in the contemporary education landscape

Tim and Martin are both affectionately known Australia-wide across industry and education sectors as the "grandfathers" of HVAC. Both trained as refrigeration mechanics in the late 1960s and early 1970s before becoming trade teachers in the 1980s. They have witnessed the restructuring of the VET sector, the political and cultural devaluing of trades education, and the impact of curriculum changes on trade apprentices, tradespeople and teachers. They have also been at the forefront of a sector that has been forced to adapt to massive technological change. Their experiences reveal both the extent of systemic decline and their emergence as crucial but unheralded resources: while witnessing the sector's degradation, they've simultaneously developed strategies to maintain educational quality. Like William, Edward, Noel and Neil, they are committed to building a better learning environment to foster new generations of fridgies and their skillsets.

One key tension of the contemporary VET landscape is the challenge of adapting the rigid national skills framework to emerging training needs in a severely underfunded vocational sector. While vocational education has established nation-wide networks to connect industry needs with student outcomes, in practice, we heard of significant lags between identifying industry needs and securing approvals for updates to specific competency or skills packages. As Tim describes, training for HVAC&R is stuck in a moment in time:

> To this day, we are still delivering content that is from the 1970s and 1980s and doesn't exist anymore ... And there's a lot of new things, yet we're not talking about [it], we're not assessing.

Lack of investment in providing time for innovation, resources for programme development, and challenging bureaucratic processes were all cited as root causes for these lags. An outdated and overcrowded curriculum has left very little scope within existing teaching roles for innovation, capacity to incorporate new technologies, or to develop programmes and skillsets which specifically target energy efficiency. As the Certificate III has become increasingly overburdened with content and outdated competency packages, teachers like Neil also raise concerns about meeting VET sector compliance requirements:

> I don't think there is a package in Australia that would be compliant to what [the compliance body] wants. This [incoming training package] will be compliant, and as far as we're concerned, between our four institutes, it's going to be as close as we could ever be.

The implication of programmes with outdated, overburdened, or non-compliant content is that it limits capacity for teachers in the VET sector to develop specific skillsets. In the case of HVAC&R, these include skills targeting energy efficiency in the built environment. Delays in addressing skills development risk stalling the momentum needed for energy transitions in the built environment.

HVAC&R teachers also cited poorly formed mentorship and support networks as another limiting factor. Across Australia, increased competition between VET institutions has devalued trade training networks. Teachers at different institutions work in silos, unable to access and share resources or the range of materials and resources they require to teach students to a high standard. As a result, HVAC&R trade education lacks consistency. According to Tim, trade teachers develop teaching and assessment material independently, even within their own institutions, based on their own interpretation of the national framework. Teachers rarely get formal opportunities to collaborate on curriculum development, leading to potential curriculum gaps, and a lack of consistency between different trade providers. Tim explained:

> My biggest bugbear through my life as an educator, and the biggest shock, was when I discovered that every teacher had written their own tests, every teacher used their own delivery content. There was 24 teachers [in Australia, at that time], and no one shared anything ... What really frustrated me, was that depending on the teacher's personal experience, that seriously governed what you learned in that topic.
>
> There were teachers there that ...were fixing domestic refrigerators – so when they were teaching supermarket cabinets, they had no experience at all. So students got virtually no coverage in that area. And then when it came to doing a test, what ... they [other HVAC&R teachers] did was totally different to the tests that I did. So there was an anomaly, there was no consistency.

Likewise, Martin describes the approach used to generate teaching material is time consuming and ad-hoc. Teachers use what they can to support their teaching, and how they think it might best fit competency outcomes:

> … we find now that probably half of half of the class haven't got [textbooks]… so we [lecturers] have a bundle of old editions they can distribute, and they just work around it and find something else on the web to support their teaching … people will use what they have available. And they tried to write competency statements that sort of matched with the workbooks they have. Now that we're getting audited to the extreme about competency, that's become a disaster. Because there was never competency in the first place.

For HVAC&R teachers, decisions made over successive rounds of policy reforms are difficult to unpick. Growing skills demand, existing workforce shortages, new technologies and changing industrial and environmental conditions all highlight the need for strong and supportive networks within the VET sector, to ensure that the right knowledge, skills and responsibilities for updated skills development can be distributed.

As Jesse Adams Stein has argued elsewhere, the devaluation of HVAC&R trades education sidesteps what the trades-based education entails in very practical terms: trades education is a "hands-on process" that is also "material, practical and embodied" and "'gained through the cumulation of hundreds of hours of exposure to wide-ranging practical tasks as well as theoretical knowledge about materials, technologies and processes."[32] This emphasis on embodied learning through practical experience stands in stark contrast to current training approaches. Tim identifies competency-based assessment as particularly problematic, taking apprentices away from developing "know-how" and skill with their hands and their tools:

> I strongly believe that we need to stop pushing young people to learn things that they don't need to know right now. You only learn something because you've been given a need to know … I spent my life (pauses) trying to provide a reason why we were doing something. Often it was because it was in this damn bit of paper, this stupid standard. I don't agree with it. It's from 40 years ago and I shouldn't be teaching you this crap, but look, I won't push too hard with the assessment. And you find yourself doing that because you know that you're ramming something down their throat that they don't need to know, and they never gonna use, which all falls apart. I'm not sure where to go from there … *The focus needs to be back on the skill. So show me you can do it and I'm happy.*

Tim's frustration points to a tension in contemporary trade training regimes: the gap between competency frameworks and the reality of how trade knowledge is best acquired and applied.

HVAC&R teachers' desire to "do the right thing" highlights the gap between rigid, outdated curriculum standards and their sense of emerging knowledge and skills needs of apprentices entering a sector of rapid change, new technology and increasing demand for built environment maintenance in a climate changing world. Their summary view is of a fragmented skills landscape. Teachers are aware that fridgies are graduating with limited breadth and depth of knowledge, unevenly distributed competencies, and with less knowledge and capability than in previous decades.

In an increasingly resource-constrained and degrading environmental context, and amid a changing landscape of technology, standards, regulation and expectation, fostering HVAC&R trade education and skills development is more critical than ever. Meeting this demand will no doubt be difficult, and future transitions will inevitably be imperfect and time-lagged. However, our work with the education and training workforce in the HVAC&R sector suggests that there is also a substantial opportunity to work with the existing knowledges, skills and know-how already embedded in the VET sector to navigate climate-changing futures. The following section positions the adaptive capacity of the HVAC&R teachers as a resource for future workforce transitions. We consider how current skills development is navigated in the context of uncertain futures, both within a constrained VET system and within broader social, economic and environmental structures.

Navigating communities of practice in formal sites of training

HVAC&R teachers often expressed a disconnect between their core values – combining education principles and a desire to "pass on" their own trade expertise – and the reductive forces of the VET sector. But despite these systemic constraints, HVAC&R teachers were driven to "give back to their trades" through the process of handing over their trade skills and knowledge to the next generation of apprentices and within trades-focused teaching networks. Our empirical research revealed how HVAC&R teachers actively resist both declining standards and the erosion of their trade. Far from being overcome by the degradation of VET institutions and rigid skills frameworks, they are successfully navigating the skills development required for future energy transitions.

As Martin describes, one of the most prominent strategies offered to face uncertain futures is "teach them the basics" and to equip people with the "right set of skills":

> We spend a lot of time working on training people on the principles. Because if you understand … the science of refrigeration … you understand the principles, and you apply that to what you're doing, you usually work out what's going on. The principles never go away. It's applying the principles is always the most important thing we can do. That's why we spend so much time on it.

Rather than understanding a distinct work process or proving yourself capable through "checkbox" competencies, Martin prioritises nurturing the "right set of skills" to enable flexibility and an understanding of how to approach non-routine tasks alongside the basics.

According to Tim, while the VET sector had taken focus "away from the skills-based side of it because [institutions] haven't had the resources given to them," there was also a deep commitment to – and recognition of – basic skills. HVAC&R teachers embraced a narrative of "quality." Rather than ensuring timely completions or increased student numbers (concerns driving the VET sector), teachers are more concerned that students actually develop the skills to work independently. This knowledge and skill-base will be crucial in addressing the overlap between the scale of workforce demand looming for this trade and the inevitable timeline in which the sector adapts.

Other strategies employed by HVAC&R teachers involved developing and strengthening networks and mentoring opportunities amongst teachers themselves. One example is a self-organised national network of HVAC&R teachers, supported by some of the "grandfathers" of HVAC&R – like Tim and Martin. In the absence of professional development opportunities across trade institutions, the network brings HVAC&R teachers together nation-wide to build collegiality, socialise and share information. As Tim describes, attempts at building a community of practice via formal sites of education has been challenging:

> … we were in a bit of a difficult situation. [VET institutions] around the country are cutting back on the ability for teachers to get together [for professional development]. We had been meeting generally twice a year for about 20 years, usually one person from each [VET institution] around the country … and it got to a point, I mean, in 2015 or 2014, where the [VET institutions] were saying, 'no, we're not paying to send you because it's costing money.'
>
> So I went to [a Peak Organisation] and said, could [you] support the attendance …to try and build consistency and travel on the same path. And he said yes. So for years now, he's been paying for the flights and the accommodation for one person from every RTO [Registered Training Organisation] to come together. I'm very proud of the people that I work with.… They're all keen to, to achieve the same goal.

These challenges emphasise the importance of recognising the social context embedded in formal sites of training, a key dimension of a "communities of practice" approach. The national HVAC&R teaching network reveals how peer collegiality helps overcome sector fragmentation while renewing and reviving trade knowledge curriculum.

This model has also filtered down to state-based networks. Teachers like William, Edward, Noel and Neil work to overcome the fragmentation of the sector, uniting across metropolitan and regional VET institutions to ensure consistency in programmes. In the process, they've strengthened connections with industry networks to close technological gaps for new equipment. As Neil describes:

> I think it's important…having that industry connection. We've been a little bit lucky … because about six years ago we made this connection with industry. Their whole concept was, the industry needs to put money in to get a better product out, and that's what we're aiming for now. By the four of us all getting together, I think we're going to achieve an even better outcome now, because we're working in a lot closer, where we were all working individually before. When the big companies like Daikin and Fujitsu and all those mobs see what we're doing now, I think they'll be even more interested in coming to the party and helping us out.

Through these self-organised initiatives, HVAC&R teachers have become increasingly important middle actors in preserving and advancing trade knowledge. Their Wednesday evening meetings, conducted after hours and without institutional support, reveal both the systemic failure of the VET sector and teachers' determination to maintain educational quality. By building and maintaining informal networks at both a national and state level, groups of teachers are working together to enhance the trade curriculum. These examples of communities of practice in formal sites of training, highlight how HVAC&R teachers are both absorbing and re-distributing knowledge and skills from the middle-out to better position themselves to deliver necessary skills to the workforce.

Conclusion

HVAC&R systems will continue to be vital infrastructures for supporting life on an increasingly volatile planet. Installing, repairing and maintaining these systems demands a robust and skilled workforce. Questions about how best to meet future skills demands cannot be separated from debates about the future of work itself. Central to this is the work of the climate crisis, and how future energy transitions are deeply dependent on a skilled workforce. It is here that attention to vocational educators – such as HVAC&R teachers – is so vital. This chapter has focused on the largely underappreciated and undervalued work of HVAC&R teachers, revealing their resourcefulness in navigating existing gaps and tensions around training. In a positive development, in February 2025, the Federal Government passed legislation providing financial assistance to states for free VET places, signalling broader recognition of the pressing need to fill critical gaps in Australia's technical workforce skills. As the implications of this policy shift unfold, the vital role of HVAC&R teachers needs to be better recognised and supported.

HVAC&R teachers are middle actors, uniquely positioned at the crossroads of existing offerings in vocational education and training, and the future needs of this in-demand trade. To date, this view has largely been overlooked, yet it constitutes a crucial resource for shaping the vital skills development pathways needed to address the energy transition. Examining HVAC&R teachers' experiences within the contemporary Australian vocational sector reveals crucial insights about the knowledge, skills and resources needed to meet both workforce demands and climate challenges across the built environment, agriculture and supply chains. Our assertion here is that the knowledge and resourcefulness of HVAC&R teachers can play a transformative role in bridging gaps and coping with lags as future transitions take place.

Our focus on HVAC&R teachers also highlights the importance of networks and relationships to bring together "precarious worlds of learning and work."[33] Recognising communities of practice in formal sites of education and training opens up the micro-spaces of existing training cultures – revealing the new knowledge, capacities and capabilities needed to navigate uncertain futures. Our research also reveals the centrality of networks across formal sites of training to nurturing appropriate skills development and institution building, and the importance of anchor organisations – such as peak organisations and national and state-led education agencies – to support these networks.

While the resourcefulness of HVAC&R teachers in navigating current gaps and tensions offers a sense of hope that the future of trades training is in the right hands, questions remain about how sustainable the self-driven "patching" of an inadequate system is. HVAC&R teachers are also confronted with larger questions and concerns regarding the future capacity of the HVAC&R training workforce itself. As the "grandfathers" retire and pass on, who will take their place? Will new teachers have the depth of trade knowledge and expertise to navigate future educational, technological and industrial transitions? The current figures for teacher recruitment are discouraging. All HVAC&R educators spoke of difficulties recruiting new staff to trade teaching: there is more money to be made "on the tools," while the contemporary Australian VET sector provides little certainty for continuing employment, high teaching workloads, even higher administrative loads, and a lack of resources for teaching delivery. While HVAC&R teachers support each other, curriculum programmes also require support through mentorship networks and connections with industry to deliver the desired skills-based outcomes.

Despite workforce pressures and institutional constraints, HVAC&R VET teachers will be essential for developing training that meets rapidly evolving industry and workforce demands. As we navigate unprecedented climate challenges, HVAC&R teachers occupy a position of unique significance, bridging theoretical and practical dimensions of energy efficiency and translating abstract climate goals into tangible skills and capabilities. As middle actors at the critical junction between industry transformation and educational innovation, they offer a crucial model for meeting the energy transition: technical expertise that remains deeply grounded in practical knowledge, while continuously adapting to meet emerging challenges.

Notes

1 In Australia's VET system, a Certificate III is a nationally recognised qualification that prepares students for skilled work. In trade areas, a Certificate III is typically completed over 3–4 years alongside an apprenticeship and combines theoretical knowledge with practical skills. Across Australia, this qualification is known as the Certificate III in Air Conditioning and Refrigeration or the Cert III in Refrigeration and Air Conditioning. In this chapter, we use Certificate III to cover all course variations.
2 Stanes et al., "Understanding the Barriers to Skills Development."
3 International Energy Agency (IEA), "The Future of Cooling," 3.
4 Schupak, "How to Make Air Conditioning Less of an Environmental Nightmare."
5 Stanes et al., "Understanding the Barriers to Skills Development," 11.
6 See also Ramsarup, McGrath, and Lotz-Sisitka, "Reframing Skills Ecosystems."
7 Stanes et al., "Understanding the Barriers to Skills Development," 11.
8 Mills, "Geographies of Education, Volunteering and the Lifecourse," 103.
9 Parag and Janda, "More than Filler," 103.
10 This research focused on "mid-tier" buildings, comprising 85 percent of Australia's commercial building stock. These typically smaller, older buildings are characterised by outdated design, poor thermal performance, and inefficient technologies. HVAC&R systems in the mid-tier account for approx. 40 percent of building energy use. Improving performance is notoriously challenging due to cost-sensitive owners and tenants favouring low-lost maintenance over best practice. See Daly et al., "Extending Urban Energy Transitions."
11 The interdisciplinary engineering and social sciences team is worth highlighting here, recognising that energy efficiency is more than a technical challenge, where work practices, policy settings, economic constraints and knowledge gaps all play a role in locking in high energy consumption.
12 Phillips, "Blue Collars, Green Jobs?"
13 IEA, "Net Zero by 2050," 17.
14 Patchett and Mann, "Five Advantages of Skill," 26.
15 Carr, "Repair and Care," 234, emphasis in original.
16 Carr "Repair and Care," 234; Stein, "Unresolved Tensions in Green Transitions."
17 Carr and Gibson, "Geographies of Making," 308.
18 Carr, "Repair and Care," 231.
19 Brown, "Hāth se Sīkhna."
20 Brown, "Hāth se Sīkhna," 1.
21 Brown, "Hāth se Sīkhna," 2; Mills, "Learning"; Lave and Wenger, *Situated Learning*.
22 Lave and Wenger, *Situated Learning*.
23 Krzywoszynska, "What Farmers Know," 293; Patchett and Mann, "Five Advantages of Skill."
24 Brown, "Hāth se Sīkhna"; Krzywoszynska, "What Farmers Know."
25 Brown, "Hāth se Sīkhna," 2.
26 Parag and Janda, "More than Filler."
27 Brown, "Hāth se Sīkhna."
28 Patchett, "The Taxidermist's Apprentice"; Stein, *Industrial Craft in Australia*.
29 See also Stein, *Industrial Craft in Australia*.
30 Frost-Camilleri, "You Can't Un-ring a Bell," 11.
31 Nakar and Olssen, "The Effects of Neoliberalism," 998; Stein, *Industrial Craft in Australia*.

32 Stein, *Industrial Craft in Australia*, 70.
33 Ramsarup et al., "Reframing Skills," 102836.

Bibliography

Brown, Trent. "Hāth se sīkhna: Geographies of Practical Learning and India's Agricultural Skills Agenda." *Third World Quarterly* 45, no. 4 (2022): 734–52.

Carr, Chantel. "Repair and Care: Locating the Work of Climate Crisis." *Dialogues in Human Geography* 13, no. 2 (2023): 221–39.

Carr, Chantel and Chris Gibson. "Geographies of Making: Rethinking Materials and Skills for Volatile Futures." *Progress in Human Geography* 40, no. 3 (2016): 297–315.

Daly, Daniel, Chantel Carr, Matthew Daly, Pauline McGuirk, Elyse Stanes and Inka Santala. "Extending Urban Energy Transitions to the Mid-Tier: Insights into Energy Efficiency from the Management of HVAC Maintenance in 'Mid-Tier' Office Buildings." *Energy Policy* 174 (2023): 113415.

Frost-Camilleri, Liam. "You Can't Un-ring a Bell: Teaching amidst VET Neoliberalist Ideology." *Fine Print* 45, no. 3 (2022): 11–13.

International Energy Agency (IEA). "The Future of Cooling: Opportunities for Energy-efficient Air Conditioning." *IEA* 2018. https://www.iea.org/reports/the-future-of-cooling#:~:text=Growing%20demand%20for%20air%20conditioners,costs%20at%20the%20same%20time

IEA. "Net Zero by 2050: A Roadmap for the Global Energy Sector." *IEA*, 2021, 17. https://iea.blob.core.windows.net/assets/deebef5d-0c34-4539-9d0c-10b13d840027/NetZeroby2050-ARoadmapfortheGlobalEnergySector_CORR.pdf

IEA. *Skills Development and Inclusivity for Clean Energy Transitions*. IEA, 2022. Accessed 8th May 2023. https://www.iea.org/reports/skills-development-and-inclusivity-for-clean-energy-transitions

Krzywoszynska, Anna. "What Farmers Know: Experiential Knowledge and Care in Vine Growing." Sociologia Ruralis 56, no. 2 (2016): 289–310.

Lave, Jean and Etienne Wenger. *Situated Learning: Legitimate Peripheral Practice*. Cambridge: Cambridge University Press, 2021.

Mills, Sarah. "Geographies of Education, Volunteering and the Lifecourse: The Woodcraft Folk in Britain (1925–75)." *Cultural Geographies* 23, no. 1 (2016): 103–19.

Nakar, Sonal and Mark Olssen. "The Effects of Neoliberalism: Teachers' Experiences and Ethical Dilemmas to Policy Initiatives within Vocational Education and Training in Australia." *Policy Futures in Education* 20, no. 8 (2022): 986–1003.

Parag, Yael and Kathryn Janda. "More than Filler: Middle Actors and Socio-technical Change in the Energy System from the 'Middle-out.'" *Energy Research and Social Science* 3 (2014): 102–12.

Patchett, Merle. "The Taxidermist's Apprentice: Stitching Together the Past and Present of a Craft Practice." *Cultural Geographies*, 23, no. 3 (2016): 401–19.

Patchett, Merle. "Historical Geographies of Apprenticeship: Rethinking and Retracing Craft Conveyance over Time and Place." *Journal of Historical Geography* 55 (2017): 30–43.

Patchett, Merle and Joanna Mann. "Five Advantages of Skill." *Cultural Geographies* 25, no. 1 (2018): 23–9.

Phillips, Leigh. "Blue Collars, Green Jobs? The Breakthrough Institute." The Breakthrough Institute, 2021. https://thebreakthrough.org/journal/no-15-winter-2022/unions-labor-green-new-deal

Ramsarup, Presha, Simon McGrath and Heila Lotz-Sisika. "Reframing Skills Ecosystems for Sustainable and Just Futures." *International Journal of Educational Development* 101 (2023): 102836.

Schupak, Amanada. "How to Make Air Conditioning Less of an Environmental Nightmare." *The Guardian*, 3 September 2021, https://www.theguardian.com/environment/2021/sep/03/air-conditioning-climate-crisis-clean-tech-solutions

Stanes, Elyse, Chantel Carr, Pauline McGuirk, Daniel Daly, Matthew Daly and Inka Santala. *Understanding the Barriers to Skills Development for Energy Efficiency in Commercial Building HVAC*. Wollongong: University of Wollongong, 2022. https://betterwaystowork.com.au/wp-content/uploads/BWW_Skills-and-training.pdf

Stein, Jesse Adams. *Industrial Craft in Australia*. London: Palgrave Macmillan, 2021.

Stein, Jesse Adams. "Unresolved Tensions in Green Transitions: Retraining and the Question of 'How'?" *Dialogues in Human Geography* 13, no. 2 (2023): 244–48.

4

QUIET SUSTAINABILITY

Tacit Knowledge and Emic Capacities in Building and Construction

Ben Lyall and Sarah Pink

Introduction

Construction industry advocates in Australia often frame the sector as a national priority: part of a nexus that includes infrastructure, real estate, employment and superannuation.[1] Media attention paid to the economic fall-out from the Covid-19 pandemic – "a string of high-profile insolvencies" – exemplifies this focus.[2] While valid, these financialised narratives have a chilling effect on longer-term discussions including the sector's outsized contribution to Australia's emissions. Materials, transport, construction and demolition works, and ongoing maintenance all draw significant energy and generate waste.[3] Local projections of the industry's future contain some sensitivity to the wider climate crisis.[4] But for those working in management roles (across residential, non-residential and engineering), climate is rarely viewed at a systems level: it tends to be articulated through singular issues, such as coping with extreme weather or monitoring waste, with responses articulated around productivity needs and policy pressures.[5]

Globally, economic framings of climate issues are dominant in the construction sector, whereby sustainability evaluations emphasise financially strategic motivations and recommendations.[6] Technological change is a popular vehicle for such strategy, and while "digital transformation" or "Construction 4.0" may *feature* sustainability, these terms largely reflect an "engineers paradigm" within (and adjacent to) the sector.[7] These perspectives imagine day-to-day work in construction "transitioning" rapidly through digitalisation and automation, but implementation is disconnected from climate issues as experienced by a majority of the industry's workers.[8]

DOI: 10.4324/9781003512240-6

We argue that such approaches enervate rather than enliven future possibilities. For manual trades, where knowledge is strongly tied to individual embodied labour and material skill, we see important questions arising in terms of how a human workforce might work more sustainably in climate crises. Returning to Australia, our chapter draws from a project focused on the current and future lives of individuals working in building and construction. Existing research in the construction industry (particularly regarding health and safety) has identified significant practical innovation at the level of individuals and teams, which are often overlooked drivers of change in the industry.[9] We likewise found individuals were cognisant of wider industry pressures and key debates regarding sustainability, and addressed their concerns through personal practices. To that end, this chapter takes an in-depth approach and draws on two interviews: one with an independent designer/builder and another with an educator at a vocational education institution. Both interviewees channel decades-long expertise into contemporary concern for sustainability.

To analyse these cases, we borrow the term "quiet sustainability" from socio-agricultural work.[10] The term originally referred to environmental benefits nested within the social outcomes of communal gardening but has since been used to generally describe micro-level sustainability mobilisation, often where regulatory structures are ambivalent or absent.[11] In adapting the concept to the construction industry, we focus on "quiet" practical engagements with sustainable practices that contrast the "loud" rhetorical messages of industry leadership. In addition, we align with existing research into safety and risk management practices on construction sites and the concept of "quiet safety," which similarly highlights the value of tacit knowledge.[12] The conceptual connection, we will demonstrate, can be seen in how personal ethics and experience often exceed regulatory mandates. More generally, we invest in the richness of first-hand accounts that – contrary to top-down industry perspectives – derive *hope* from individual practices rather than *hype* from technological promises.[13]

Project context, methods and participants

This chapter draws from interviews conducted as part of the AUTOWORK project, which explores work futures through digitalisation, automation and robotisation. The project is ethnographic and explores three sectors (retail services, healthcare, construction), with the aim of foregrounding individual worker perspectives. In the construction stream of the project, we built on past short-term ethnographic work and sought diverse perspectives on future opportunities and challenges.[14] We interviewed and engaged with the everyday working environments of participants using visual ethnography methods. Within the ad-hoc contexts of businesses operations or projects, this included video recorded interviews, on-site walking interviews in workplaces such as construction sites and start-up workshops (using video and photography where agreed), and remote video interviews using digital platforms. Using video in ethnographic research enables participants to better show, demonstrate and engage researchers in their everyday practices, experiences and environments. Participants (totalling 20) included independent builders, surveyors, safety officials and managers in large companies, educators and technologists.

In this chapter, we focus on two on-site interviews (both conducted in the Australian state of Victoria) that offer valuable insights into how practices might be reshaped to meet a precarious climate future. We quote participants at length. First, because their discussion captures spaces and gestures (acting in lieu of audio-visual material), and second, because both demonstrate an "energy literacy" that previous research has indicated is required to progress the industry's sustainable practice.[15] Interview one is with Eugene, who has been working as an independent designer/builder for almost 30 years. His company focuses on environmentally sound approaches to both renovations and new builds. At the time of interview, Eugene was working (alongside an apprentice) on a house renovation in Melbourne's inner-northern suburbs. Interview two was conducted at a large regional vocational education institution, with experienced teacher Mark. Mark has worked in education roles for 20 years, after a previous 20-year career as a second-generation cabinetmaker (originally an apprentice to his father). Despite doing very different work, both Eugene's and Mark's cases are emic examples of "quiet sustainability" in building and construction.

Eugene: Active work on a passive house

I met Eugene (man, 50s) on-site, and we toured the bare bones of the building. The renovation being undertaken is extensive: the house is gutted, and an additional storey is emerging. The end-goal is a highly efficient modern home, specifically, a "passive house." Based on the German *passivhaus*, this design methodology emerged in the 1970s and was codified as an ISO (the International Organization for Standardisation) standard in the 1990s.[16] As Eugene explained:

> … it's a German building standard from the '80s. An interested scientist started working through building physics to make use of the least energy that they could. It's a cold country! So now it's the main standard around the world for low-energy building. It's based on an energy-use figure which is the same anywhere in the world. You don't have to get it perfect, you just have to beat that figure to get the house to a level where it can be certified. Then they've got a few different levels of that. The figures they use are that it will use 10 percent of the energy of a normal building, a normal leaky building.

> They've got five key criteria that they use to differentiate from a normal building, but you can even do the economic modelling. When you do the course – if you want to keep studying on your own and working through the numbers – you can present [to clients] the energy costs: this is how much this is going to cost you extra, and this is how long it'll take you to pay that back. They [the *Passivhaus Institut*] tried to just sort everything out so you can't really argue against the system. Lots of people think we don't need it, but the numbers have all been done [and] it will value-add to the house if the people sell it, it's proven.

The criteria Eugene mentions relate to: thermal insulation, minimal thermal bridges (materials that carry across a building's internals and external facades), air-tightness, windows and doors that are well sealed and appropriately placed to control temperature seasonally, and mechanical ventilation that exchanges stale internal air for fresh air (with the additional possibility of recovering humidity and heat from the former).[17] Indeed, the ventilation system being woven through the old and new elements of the building's wooden frame is a striking element of the project.

The performance criteria for *Passivhaus Institut* certification are fixed, and require specific design and building knowledge. Eugene explained how the "trade" and "design" courses associated with the certification interact, from his perspective of having completed both: "the tradies [tradespersons] need to know how to put it together, and the designers need to work through all the physics [and the computer] programs." Eugene undertook *passivhaus* training in 2019 and a subsequent "Cert[ificate] IV in Energy [Efficiency and Assessment]" during the pandemic. Eugene outlined the practicalities of implementing *passivhaus*, in a personal career context where he had seen many flawed buildings:

> You can use lots of different building methodologies but you are still basically ventilating the building; making sure that it's airtight and waterproof; no thermal bridging. The fundamental rules are there but you can address it in different ways. It is a steep learning curve for most people when they've just been using standard methods. Most of the architects on the course, a lot of them struggled. Engineering people probably find it easier. Architects tend to be a bit more creative, like visually, thinking about how the thing is going to look and how to move through it, rather than the building physics. It's definitely a step up for tradies and designers.
>
> But once you've done it – like a lot of things, once you know it – then it kind of seems obvious [chuckles]. You know what I mean? … Using wrong materials. We've been building wrong for so long! We've been putting foils on the outside of the building that lock all the moisture in it. Then you've got a building that's not airtight on the inside, moisture can just get into the

> walls and it's trapped in there; it can't breathe. But yeah, with this blue wrap, it's vapour-open. It's waterproof but if there is moisture in the wall, as the heat pushes through, it will allow it to just evaporate. So it's like a safety shield on the outside. Then the airtight layer on the inside. So we've been putting these foils on, insulating walls, and putting cladding hard on that, so it's all just locked up.
>
> We've all seen the articles over the years of leaky buildings and stuff. But it's a bit more than that, it's health as well: mould, asthma. A lot of people are susceptible to asthma just from black mould growing on the walls and the spores just going everywhere. I've seen a lot of it over the years, during demolitions.
>
> [Owners] want to reduce the footprint and they want a good performing house … but you actually don't know what you're getting in the building performance whatsoever. The star rating [an existing energy rating system for new buildings] is completely pointless, almost. It's okay on paper but it's not policed and has zero follow-up. You've got a six star on paper and that's it; 'this is endorsed.' You can't hide in this [gesturing to the surrounding building]. If you don't build it and tight enough, then you failed.

Eugene's disappointment in Australia's building standards is palpable. He notes that contemporary "buildings are pretty highly insulated," but other aspects of building codes "haven't really stepped up." Since the *passivhaus* standard was developed in Europe, it has no formal place in the Building Code of Australia.[18] As alluded to by Eugene, building codes in Europe are more stringent, and application of the principles in Australia relies on the resources of individual builders and building owners.[19]

Eugene explains how the pursuit of *passivhaus* emerged in the current project. The client was eager to integrate renewable energy during renovation, but Eugene identified how materials in the existing heritage house could undermine such goals. Thus, Eugene advocated for *passivhaus* certification, so that the renovation would meaningfully reduce energy usage *and* costs in the long term. In many ways, the project seems to represent an ideal convergence of Eugene's expertise and interests, but as a "pretty small-scale business" he remained circumspect about each project, particularly on renovations like this where materials pose challenges:

> That's a trick, renovations! This one is stripped right out. A lot of people will just baulk at it because you need to get the whole building up to scratch. Some people try doing two: the old part and the new part, working differently. But if you don't fix the old part up, the ventilation system is properly struggling because it's so leaky. It's a bit of a waste.
>
> So for instance, this job, it was me that drove it. I came and looked at it and it's a really hard site to do a design that would work with passive solar [an

interest of the client], because this old house is heritage listed and had to stay. So I knew about the system. I knew about it pretty much when the first guys in Australia started. But I was a bit like, 'why would you bother here,' because of our climate. But you know – we've all, in the last 10 or 15 years – our awareness of energy use has just become so much more focused.

You're always thinking of costs. There's obviously been an increase across-the-board in all labour and materials and a lot of builders are on fixed price contracts like I am, so you know you've lost potentially quite a bit of profit on a job. … So basically you'll be working for less. Then if you take on jobs that are harder for you, and probably haven't been able to estimate them accurately.

It's been a journey. We had, obviously, issues with supply chains and just the whole building industry with orders getting cancelled and trying to find people. It's pretty crazy at the moment. I get this kind of feeling – apart from the war [in Ukraine] and energy costs, that have definitely had a big effect on our inflation – that when this happened, it's like we're paying closer to the real price. I just feel like – especially in Australia – we've been so fortunate with resources, mostly timber. It's so scarce in the rest of the world. I just figured it was a little bit more of a balance-up. … I didn't realise how much of our stock was imported engineered timber. I knew the big companies were, but yeah, most of it is just actually engineered in the States and Canada and shipped here. … I had to go to non-engineered solid timber and things like that, some steel beams, because we couldn't get the other stuff. … So yeah, you're always learning.

Eugene's reflections on the intersections of project-specific material needs, local market patterns and global geopolitical stressors (armed conflicts and the trade distribution impacts of the Covid-19 pandemic), are widely demonstrative: fallout from specific crises are often plural, as are the current and coming impacts of the climate crisis. Following the aforementioned tendency of construction industry management to compartmentalise risk,[20] we again emphasise individual expertise – "always learning" – in changing everyday practices.[21]

While acting in concert with broader technological shifts, construction methodologies like *passivhaus* are animated by material practices – not just a "solutionist"[22] approach to retrofitted smart ventilation systems[23] – and embodied by people like Eugene. Issues arise (and will continue to) when materials are scarce, but Eugene is also exploring low-energy materials like rammed earth and hemp for future builds. Depending on the project, Eugene calls on either or both of his design and building expertise, and in each case, shares elements of his sustainability interests to those working with him. Eugene's practices are a reminder – particularly with his current renovation only a few blocks away from the concrete pouring crane of a nearby high-rise apartment project – that "people who are skilled in dealing with the material world ... offer a powerful challenge to the idea of the industrial city as terminally ill or lacking resilience."[24]

Mark: Manufacturing a new generation

I met Mark (man, 60s) at the regional vocational education institution where he is employed. The building is mixed use: two-storey classrooms eventually open out into double-height workshop spaces, then outdoor areas. These spaces are abuzz with teaching and learning of building trades and skills including formwork, bricklaying, carpentry, scaffolding, roofing, and rigging. Weaving between various table saws, routers and jigs to find a quieter area, we discussed Mark's interests in efficient and sustainable carpentry, and how institutions should function effectively to equip future generations. Working in an educational setting, Mark contrasts with Eugene, insofar as his interests are expressed through institutional goals, rather than project-based praxis or apprenticeship.

During his time at this institution, Mark has travelled overseas to observe emerging trends. In 2011, he was awarded a fellowship by the *International Specialised Skills Institute* and visited several European factories to observe cabinetry manufacturing processes. He authored a report for Skills Victoria[25] (a former state government body) based on this work, and continues to "keep up to date on the internet about what's happening." Mark saw much room for innovation in Australia, particularly in managing hazardous waste from composite products like MDF (medium-density fireboard):

> In Europe they'll burn all the waste. They have large furnaces that are highly efficient. They burn [waste] in chambers, so there's very low emission and they use that to produce energy to run their factories or heat their factories. ... it can be burnt but you need an enclosed system to actually burn it.
>
> They do recycling of materials but they're close [nearby]. When they build their factories they built it near the guy that makes the board. Tyranny of distance – cost – that's the problem with waste here [in Australia]. I think we've got two board manufacturers, one in Ballarat and one in Queensland. So to get waste back to those places you chew up more in carbon. ... We just send bins full of board to landfill so there's nothing else you can do with it. People can't burn it domestically, it's got chemicals in it.
>
> I've seen factories where – for housing – it's all done on CNC [computer numeric control]. They produce all the frames and then they get loaded onto a truck. I don't think they're using CNC so much in frame production here, not like they do in Europe. They produce whole panels of sides of houses and all this sort of stuff ... no storage, straight onto the container, straight to the customer, and no packaging ... off they go and not one piece of paper is produced.

Mark admired the planning underpinning production in Europe, and lamented that the "tyranny of distance" in Australia makes this difficult. While Mark pointed to a nationwide problem, the issues played out in the microcosm of the training institution: teaching-focus aside, the spaces operate as "real" workshops and building sites, facing the same waste issues.[26]

In this regard, Mark detailed a repeated request for the institution to purchase a "briquette machine." In comparison to other teaching areas – such as bricklaying, where projects can be constructed and demolished repeatedly – the materials involved in joinery, including wood and composite materials, generate more waste. Indeed, suction equipment for particulate dominated the building's ceiling spaces. Mark outlined the difficulties inherent in such requests, with reference to briquetting "clean shavings." He sees Europe's more systematic approaches to composite material as ideal:

> When we machine up material it goes into shavings. You can compress all that [with a briquette machine] and it could be turned into heat bricks to burn. I've put in submissions before, but they won't do it. For a large skip you'd probably fill two of those a year if we compressed everything, against probably emptying a skip every three weeks.
>
> The only time we can get new machinery here is to show that it's more advanced technology. So that's the only way we get grants. A couple of [other] machines here are getting a bit older like the edger. We had a guy service it, 'oh you're still got this old thing,' and look, it's only probably ten years old I suppose, but the interface has totally changed. If there's some money floating around we might say, 'look you could upgrade that' [or] the saws, because of the safety aspect on the saws, takes the pressure off us as teachers not to have the eyes in the back of our heads all the time.

Mark recognised multiple beneficial upgrades to the workshop but expressed some frustration at the internal competition on spending. As suggested by the "service"-person interaction above, Mark also draws on trade connections to decide when to push for grants in any one area. He spoke at length about the importance of links with local industry, name-checking several local businesses and sales "reps" (representatives). Importantly, Mark did this with a clear sense of *what* local suppliers produce, *how* they produce it, and the types of machines involved. Both aspects motivate his decisions. He may have a preferred approach, but knows he must balance sustainability, efficiency, safety and training best-practice.

Among the production evolutions Mark has observed in timber-based trades, CNC was significant for hybridising work practices. Chiefly, he identified how the local labour context created an ongoing overlap between carpentry and manufacturing of various types. Over time, he observed work shifting from specialised carpentry, to a division of labour with machinist roles, to the present forms of cabinetmaking: he described this as "knocking together" prefabricated elements. Like some of his contemporaries in adjacent trades, Mark did not see technology exclusively as a driver of decline.[27] Rather, he described a "natural attrition" in some occupations, somewhat counteracted by gains in efficiency: "minimal waste on your sheet when the machine's cutting it out," and reduced skill barriers (which made some workers employable across multiple types of production):

> Some of them [new students and apprentices] would have been in production jobs anyway, in [automotive manufacture] and those sort[s] of places which don't exist anymore … In the old days you'd have a machinist do all the material for you, and then it'll come to your bench and you put it together basically. But that's all changed now. The machinery does all the components and those that run the machines, they're generally cabinetmakers now … It's not that the people are being sacked or anything but it's been a gradual thing over time, so, natural attrition I suppose you'd say.

The training provided at Mark's institution anticipates a working reality where design programs, spreadsheets, CNC cutting tools and their digital interfaces are essential to the offsite preparation of components later assembled on site:

> We start [students] off at the basics and at the end they've got to produce a folio and actually build a kitchen from start to finish. So they've got to talk to the customer, do the brief, do the costing, cutting list, timelines, OH&S, sustainability. They've got to do all of those processes and actually make the kitchen, put it together, and install it. But in the workplace they might only be putting the cupboards together, or pushing buttons on the machine, or doing installations … we try and get them to do the whole job so at least they've had experience of doing it, to benefit them.
>
> They get a fair bit of solid timber and traditional cabinetmaking, we show them a bit of that. They get to use all the traditional machines, the older machines, so they've got a bit of an idea. And things – fashions – change, you know? We show them how to make a panel door and that's probably been out of fashion for the last 10, 15 years, but it might come back, you never know. They know how to put a solid benchtop together. We teach them a few other things, broaden them out a bit … we try our best to give them some more skills, but that's just the nature of the industry. Just the way it's gone. They've just deskilled them.
>
> We like to keep up-to-date with the latest so then the apprentices see it. Sometimes they say, 'oh I've never seen one of those.' We've got to be seen to be up-to-date. You don't want kids coming here and saying, 'oh geez they're back in the 19th century' … if we've got the latest technology here, that feeds through to the apprentice.

These reflections on technology are a pivot in the conversation with Mark. His generally circumspect perspective on emerging materials and techniques was tempered. "They" ("the industry") had "deskilled" the workforce in some areas: through standardisation, technological change, prefabrication and cycles of design "fashion." Mark's tone suggested he saw traditional skills – those still partially taught, such as working with "solid timber" – as intrinsically valuable, even if they are not called upon daily. Despite the tensions and overlaps between manual and digital skills, vocational training remains contemporaneous with work practices. The training moves with technology, but perhaps most importantly, it does so in a way that maintains an integration with communities of practice (which are present through connections to local business, sales representatives and apprentices).

Mark is not alone – other educators share similar perspectives on how contiguous evolution in training sustains contemporary work practices – but is overall, optimistic.[28] Mark looks at building work as an ecosystem that moves forward through the decisions that new workers can make, drawing on their training in various material practices. For Mark, there was a sense of lateral possibilities: a building trade that comes to include other skillsets and qualifications presents new options. He noted success stories – "the ones that are going to do well" – emerging. He described a former student (also a qualified electrician) now handling kitchen and bathroom projects end-to-end. Quiet sustainability, in Mark's case, manifested in contributing his knowledge towards highly localised and hybridised forms of work, cutting across trades, craft and prefabrication, and strong connections with community.

Everyday actions of quiet sustainability

In both interviews, the industry necessity of knowledge sharing remains vital, but also carries forward "quiet sustainability."[29] Eugene and Mark touch on key issues limiting the building industry's shift to more sustainable forms of production and work in the context of climate change, organisational change and regulatory reform.[30] But vitally, they showcase the enactment of positive change through everyday action. In discussing these issues with fellow builders, suppliers, educators, apprentices (and researchers), they attest to the essential social value of energy literacies to the industry's future.[31] Contrary to top-down views on industry transformation – where sustainability is shrouded by economics – we see two highly motivated individuals making the best decisions they can in the present, and positively impacting the work of future generations.

Tacit knowledge transfer (a "hidden skill base"),[32] is a common theme in ethnographic research on construction work, and we see sustainable practices as a feature of this transfer.[33] Like "quiet safety" in construction (and "quiet sustainability" in small-scale food production), Eugene and Mark's "quiet" sustainability practices are amplified through social networks. As a final testament to the spread of sustainability values through emerging approaches, Eugene described borrowing an associate's thermal camera to assess building performance, while Mark recounted helping local businesses to access equipment at the training institution. Values are disseminated in incidental moments like these reflecting that – in addition to tools and technologies – work in this industry is reliant on physical/material practices, emplaced knowledge, and wider organisational understandings.[34] Eugene and Mark are clearly agents of change, and draw on decades of experience. The resources they possess are not universals of the industry. Elsewhere, participation in sustainable building programmes have been underwhelming,[35] and the lack of interoperable industry information about climate change is a concern.[36] Indeed, Eugene's efforts demonstrate work well beyond current building standards, and from Mark we learn about institutional pressures that stymie progress on sustainability.

Fortunately, Mark and Eugene invest in developing their knowledge in ways that not only enhance the work they do, but the work others might do in the future. With the building industry comprising a spectrum of occupations, experiences and employment types (often negatively denoted as "fragmented"),[37] orientations towards change are inherently tied not just to what is possible, but what is practical. Yet they suggest higher-order values are held in the industry about doing more with less. Such an ethos provides a much sharper rendering of the work required to meet climate-related challenges, than what is seen in "transformation" focussed industry projections. Regardless of the statements, claims and agendas of the industry's larger players, meaningful change quietly hits the ground through agents like Eugene and Mark.

Notes

1 Property Council of Australia, "Australia's Property Industry."
2 Jurss-Lewis et al., "Fears More Construction Companies Will Go Broke."
3 Adey et al., "Just Transitions in Australia," 66–67; Yu et al., "The Carbon Footprint of Australia's Construction Sector," 216.
4 Renz and Solas, "Shaping the Future of Construction"; Quezada et al., "Are You Ready for Change?," 36.
5 Hurlimann, Warren-Myers, and Browne, "Is the Australian Construction Industry Prepared for Climate Change?," 131, 135.
6 Lima et al., "Sustainability in the Construction Industry"; Ortiz, Castells, and Sonnemann, "Sustainability in the Construction Industry."
7 Pries and Janszen, "Innovation in the Construction Industry," 44.
8 Schönbeck, Löfsjögård, and Ansell, "Quantitative Review of Construction 4.0 Technology."
9 Lingard and Turner, *Work, Health and Wellbeing in the Construction Industry*.
10 Smith and Jehlička, "Quiet Sustainability," 155.
11 Leach and Scoones, "Mobilizing for Green Transformations."
12 Pink and Morgan, "Short-Term Ethnography," 358.
13 Pink, "Doing Anthropology Forward," 179.
14 Pink and Morgan, "Short-Term Ethnography"; Pink et al., "Ethnographic Methodologies for Construction Research"; Brett, Thomson and Dainty, "Exploring Craft in Construction with Short-Term Ethnography."

15 Clarke, Gleeson, and Winch, "What Kind of Expertise Is Needed for Low Energy Construction?" 79.
16 ISO is a global federation, comprised of national bodies, that works with government and industry to establish standards for specific technical subjects. They set definitions for key terms, design criteria, methods for judging adherence, and requirements for maintaining a given standard. International Organization for Standardization, "Ergonomics of the Thermal Environment."
17 Dequaire, "Passivhaus as a Low-Energy Building Standard," 384–85; Brimblecombe and Rosemeier, *Positive Energy Homes*.
18 Marlow, "Passive House."
19 Hurlimann et al., "Is the Australian Construction Industry Prepared for Climate Change?" 135.
20 Yu et al., "The Carbon Footprint of Australia's Construction Sector."
21 Clarke, Gleeson, and Winch, "What Kind of Expertise Is Needed for Low Energy Construction?"
22 Morozov, *To Save Everything, Click Here*.
23 Dequaire, "Passivhaus as a Low-Energy Building Standard," 390.
24 Carr and Gibson, "Geographies of Making," 307.
25 McDonald, *Sustainability in the Manufacturing of Kitchen and Bathroom Cabinets*.
26 Navaratnam et al., "The Challenges Confronting the Growth of Sustainable Prefabricated Building Construction in Australia," 11.
27 Stein, *Industrial Craft in Australia*, 123–24.
28 Stein, *Industrial Craft in Australia*, 85, 152; Veen, Teicher, and Holland, "Continuity or Disruption?" 207.
29 Smith and Jehlička, "Quiet Sustainability."
30 Hurlimann, Warren-Myers, and Browne, "Is the Australian Construction Industry Prepared for Climate Change?" 136.
31 Clarke, Gleeson, and Winch, "What Kind of Expertise Is Needed for Low Energy Construction?" 79.
32 Rooke and Clark, "Learning, Knowledge and Authority on Site.".
33 Brett, Thomson, and Dainty, "Exploring Craft in Construction with Short-Term Ethnography."
34 Moore, "'On the Tools.'"
35 Graham and Warren-Myers, "Investigating the Efficacy of a Professional Education Programme."
36 Hurlimann et al., "Is the Australian Construction Industry Prepared for Climate Change?"
37 Renz and Solas, "Shaping the Future of Construction."

Bibliography

Adey, Peter, Sarah Pink, Rob Raven, Paris Hadfield, Adam Badger, Yolande Strengers, Darren Sharp, et al. *Just Transitions in Australia: Moving towards Low Carbon Lives across Policy, Industry and Practice*. Report. Monash University, Royal Holloway University of London, 27 July 2022. https://apo.org.au/node/318754

Brett, Richard, Derek Thomson and Andrew Dainty. "Exploring Craft in Construction with Short-Term Ethnography: Reflections on a Researcher's Prior Insight." *Construction Management and Economics* 40, no. 5 (2022): 359–73.

Brimblecombe, Robin and Kara Rosemeier. *Positive Energy Homes: Creating Passive Houses for Better Living*. Melbourne: CSIRO Publishing, 2017.

Carr, Chantel and Chris Gibson. "Geographies of Making: Rethinking Materials and Skills for Volatile Futures." *Progress in Human Geography* 40, no. 3 (2016): 297–315.

Clarke, Linda, Colin Gleeson and Christopher Winch. 'What Kind of Expertise Is Needed for Low Energy Construction?' *Construction Management and Economics* 35, no. 3 (2017): 78–89.

Dequaire, Xavier. "Passivhaus as a Low-Energy Building Standard: Contribution to a Typology." *Energy Efficiency* 5, no. 3 (2012): 377–91.

Graham, Emily and Georgia Warren-Myers. "Investigating the Efficacy of a Professional Education Program in Promoting Sustainable Residential Construction Practices in Australia." *Journal of Cleaner Production* 210 (2019): 1238–48.

Hurlimann, Anna Catherine, Georgia Warren-Myers and Geoffrey R. Browne. "Is the Australian Construction Industry Prepared for Climate Change?" *Building and Environment* 153 (2019): 128–37.

International Organization for Standardization (ISO). "Ergonomics of the Thermal Environment: Analytical Determination and Interpretation of Thermal Comfort Using Calculation of the PMV and PPD Indices and Local Thermal Comfort Criteria." *ISO*, 2005. https://www.iso.org/obp/ui/en/#iso:std:iso:7730:ed-3:v1:en

Jurss-Lewis, Toby, Shane Oliver, Tim Reardon and Adam Profke. "Fears More Construction Companies Will Go Broke." *ABC News*, 12 March 2023. https://www.abc.net.au/news/2023-03-13/fears-construction-companies-will-go-broke/102087178

Leach, Melissa and Ian Scoones. "Mobilizing for Green Transformations." In *The Politics of Green Transformations*, edited by Melissa Leach, Peter Newell, and Ian Scoones, 119–33. London: Routledge, 2015.

Lima, Luanda, Emanuelly Trindade, Luciana Alencar, Marcelo Alencar and Luna Silva. "Sustainability in the Construction Industry: A Systematic Review of the Literature." *Journal of Cleaner Production* 289 (2021): 125730.

Lingard, Helen and Michelle Turner. *Work, Health and Wellbeing in the Construction Industry*. Abingdon, Oxon; New York: Routledge, 2023.

Marlow, Andy. "Passive House." *YourHome*. Australian Government, 2020. https://www.yourhome.gov.au/passive-design/passive-house

McDonald, Mark, 2011. *Sustainability in the Manufacturing of Kitchen and Bathroom Cabinets*. Melbourne, Australia: International Specialised Skills Institute/Victorian Government. https://www.issinstitute.org.au/fellowship-report/environment-%26-sustainability/austria%2C-italy%2C-germany%2C-usa/sustainability-in-the-manufacturing-of-kitchen-and-bathroom-cabinets

Moore, P. "'On the Tools': The Physical Work of Building and Renovating Houses in Perth, Western Australia." In *Ethnographic Research in the Construction Industry*, edited by Sarah Pink, Dylan Tutt and Andrew Dainty, 58–72. Abingdon: Routledge, 2012.

Morozov, Evgeny. *To Save Everything, Click Here: The Folly of Technological Solutionism*. New York: PublicAffairs, 2013.

Navaratnam, Satheeskumar, Aarani Satheeskumar, Guomin Zhang, Kate Nguyen, Srikanth Venkatesan and Keerthan Poologanathan. "The Challenges Confronting the Growth of Sustainable Prefabricated Building Construction in Australia: Construction Industry Views." *Journal of Building Engineering* 48 (2022): 103935.

Ortiz, Oscar, Francesc Castells and Guido Sonnemann. "Sustainability in the Construction Industry: A Review of Recent Developments Based on LCA." *Construction and Building Materials* 23, no. 1 (2009): 28–39.

Pink, Sarah. "Doing Anthropology Forward: Emerging Technologies and Possible Futures." In *Anthropological Optimism: Engaging the Power of What Could Go Right*, edited by Anna J. Willow, 177–91. New York: Routledge, 2023.

Pink, Sarah and Jennie Morgan. "Short-Term Ethnography: Intense Routes to Knowing." *Symbolic Interaction* 36, no. 3 (2013): 351–61.

Pink, Sarah, Dylan Tutt, Andrew Dainty and Alistair Gibb. "Ethnographic Methodologies for Construction Research: Knowing, Practice and Interventions." *Building Research and Information* 38, no. 6 (2010): 647–59.

Pries, Frens and Felix Janszen. "Innovation in the Construction Industry: The Dominant Role of the Environment." *Construction Management and Economics* 13, no. 1 (1995): 43–51.

Property Council of Australia. "Australia's Property Industry." Accessed 14 May 2023 (n.d.). http://campaign.propertycouncil.com.au/our-campaign/australias-property-industry

Quezada, George, Alexandra Bratanova, Naomi Boughen and Stefan Hajkowicz. "Are You Ready for Change? Farsight for Construction: Exploratory Scenarios for Queensland's Construction Industry to 2036." *CSIRO*, 2016. https://doi.org/10.4225/08/58557E0B380AB

Renz, Andreas and Manuel Zafra Solas. "Shaping the Future of Construction: A Breakthrough in Mindset and Technology." *World Economic Forum*, 2016. https://www.weforum.org/reports/shaping-the-future-of-construction-a-breakthrough-in-mindset-and-technology/

Rooke, John and Leslie Clark. "Learning, Knowledge and Authority on Site: A Case Study of Safety Practice." *Building Research & Information* 33, no. 6 (2005): 561–70.

Schönbeck, Pia, Malin Löfsjögård and Anders Ansell. "Quantitative Review of Construction 4.0 Technology Presence in Construction Project Research." *Buildings* 10, no. 10 (2020): 173.

Smith, Joe and Petr Jehlička. "Quiet Sustainability: Fertile Lessons from Europe's Productive Gardeners." *Journal of Rural Studies* 32 (2013): 148–57. https://doi.org/10.1016/j.jrurstud.2013.05.002

Stein, Jesse Adams. *Industrial Craft in Australia: Oral Histories of Creativity and Survival*. Cham, Switzerland: Palgrave Macmillan/Springer, 2022.

Veen, Alex, Julian Teicher and Peter Holland. "Continuity or Disruption? An Assessment of Changing Work and Employment in the Victorian Construction Industry." *Labour and Industry: A Journal of the Social and Economic Relations of Work* 27, no. 3 (3 July 2017): 193–212. https://doi.org/10.1080/10301763.2017.1370660

Yu, Man, Thomas Wiedmann, Robert Crawford and Catriona Tait. "The Carbon Footprint of Australia's Construction Sector." *Procedia Engineering, International High-Performance Built Environment Conference – A Sustainable Built Environment Conference 2016 Series (SBE16), iHBE* 2016, 180 (2017): 211–20. https://doi.org/10.1016/j.proeng.2017.04.180

5

"CREATIVE TECHNICIANS" AND "TECHNICAL CREATIVES"

Transferable Skills for Challenging Times

Jesse Adams Stein

Introduction

In the initial provocation for the *All Hands on Deck* symposium in 2023, Chantel Carr and I hypothesised that the likely future global challenges – decarbonisation, a warming climate, geopolitical unrest – will require a significant reorientation of skill. We suggested that future skillsets will require capacities that are practical, technical, materially resourceful, social and care-based, but also creative and design-based.[1] Craftspeople, manufacturing tradespeople, and arts technicians are some of those workers and practitioners who possess essential and transferrable manual and technical skills, as well as having deep knowledge of materials.[2] In addition, they often share a commitment to more sustainable methods of production, consumption and materials circulation, making their knowledge indispensable for our fraught future.

Despite there being truth to the assertion that an increasingly deindustrialised Australia is at risk of losing its traditional craft skills base,[3] there are still some extraordinarily skilled craftspeople, arts technicians, tradespeople and creative practitioners working, making and teaching. Some of these practitioners, however, are not easily categorisable in terms of education or economic sector. Or rather, when they are necessarily categorised (for example, in Australian and New Zealand Standard Classification of Occupations [ANZCO]), some pieces of the puzzle are missing. There are some practitioners who "slip through the cracks" and whose knowledge and skills are regularly misunderstood, ignored, or simply invisible to policymakers, educational leadership and the political class.

DOI: 10.4324/9781003512240-7

The core concept that underpins this chapter is the existence of practitioners whom I refer to as *technical creatives* and *creative technicians*. I use these terms as a provocation, without the intention to affix a term for future use. *Creative technicians* and *technical creatives* refers to a loose group of workers/practitioners who – often for reasons of occupational survival – are both "creative" *and* "technical," operating cross-sectorally (that is, between industrial production and creative practice). I elaborate on this definition in the following section.

Importantly, there are broader implications to this discussion. *Technical creatives* and *creative technicians* are indicative of a larger issue that is becoming increasingly apparent: the nature of working life is shifting rapidly, and the required skills to navigate future work are likewise entering a tussle between the necessity of practical skills, and the pressure to be technologically "savvy" – not least in the face of generative AI and industry 4.0. As a result, the standard categories of skill and employment – for example those used by government agencies and the private sector – are increasingly a poor fit for people's actual working experiences. The systems and structures of organising and governing work (industrial legislation, government "skills lists," higher and further education) are no longer equipped to handle the dangerous future ahead. We therefore need to rethink skills, education and training systems and employment transition capacities, often in radical ways.

With this in mind, the example of *technical creatives* and *creative technicians* draws attention to how we account for people's *actual* skills (rather than just on-paper qualifications) – revealing some of the ways in which current understandings are lacking. In a wide range of occupational areas, there exist workers who occupy various hybrid or "in-between" roles, formally or informally. The realities of working life require a full gamut of other skills and capacities that may not, at first glance, seem related to a particular job. This might include improvisation, haptic or tacit knowledge, historical background knowledge, manual dexterity and the hard-to-define ability to just "get shit done." These sorts of skills and capacities are not easily accounted for in statistical reviews, but can be analysed through detailed qualitative research, as demonstrated in this chapter (and several others across this collection).

This chapter uncovers what I learned by talking in depth to *technical creatives* and *creative technicians* about how they navigate precarity, both in the recent past and in the contemporary Australian context. It considers the training pathways and decisions that participants made, and their reflections on those decisions. I also simply tell the stories of some of these practitioners, in the process, I hope, humanising them, demonstrating their complexity, alongside their nuanced skills and occupational positioning. In this chapter I draw from interviews in the *Makers, Manufacturers and Designers* (MMD) oral history project (introduced further on). Although in this chapter I only feature a handful of stories from MMD, my empirical findings reflect a broader set of interviews, including from my earlier oral history projects.[4]

In the section that follows, I first provide some necessary definitions and examples to provide some foothold for the concept of *technical creatives* and *creative technicians*. Following this, I justify why this hybrid group deserves our attention, and then briefly introduce the oral history project that inspired this focus. From there, the chapter shifts into the oral history material, allowing the lives and voices of four of these practitioners to come to the fore.

Ultimately, this chapter pushes us to acknowledge the hybrid, "in-between" nature of many people's skillsets, and it does this in the context of the precarious realities of working life. While my empirical examples generally come from the creative industries, the implications apply across a wide range of occupational areas and fields. Having a nuanced understanding of skill is particularly important in the context of decarbonisation and energy transitions, where those who are physically *doing the work* constantly encounter unprecedented practical problems. The way such problems are solved will often not be through neatly drawn plans or computer modelling, but on a place-specific case-by-case basis, through hands-on trial-and-error, negotiation, and through creative and technical solutions undertaken by those "on the ground."

Defining technical creatives and creative technicians

When I use the terms *technical creatives* and *creative technicians* I am referring to practitioners who possess deep technical skills in one or more technical trade or craft/making practice, and who have experience working, teaching or training across both creative arts sectors (e.g. visual arts, craft, design, music, film and television) *and* industrial sectors (e.g. manufacturing, construction). This may include people who officially classify as: tradespeople, designer-makers, artisanal craftspeople, art installers, vocational educators, arts technicians and production workers, among others.

The working and educational histories of this group tend to be complex. This cross-sectoral positioning can sometimes occur simultaneously, as practitioners take on multiple casual and contract roles. Alternatively, shifts between industries can evolve over time, as people move from one economic sector to another, bringing their knowledge with them. They may have begun their careers in a manufacturing or construction trade, but due either to precarity or dissatisfaction with the work, sought employment and training in creative fields. Some ex-manufacturing workers, for example, may have left their original trade, and turned their significant practical skill towards developing their own artistic practice.[5] Other practitioners may have initially trained in a creative field – such as product design, fashion or visual art – but increasingly found themselves involved in manufacturing production, for instance in demand by local manufacturers due to their (now rare) practical skillsets. Others may have begun their careers as "designers" in a more orthodox sense, and transitioned to self-led maker production in an effort to operate with more ecological and ethical responsibility, becoming manufacturers in their own right.

A few earlier studies touch on similar phenomena. In 2010, sociologist Mark Banks focused on under-recognised "supplementary" craft and technician roles within creative labour in the United Kingdom.[6] Read today, a decade and a half later, Banks' description of precarity seems mild in comparison to the wholesale "gig-economy" state of so many industries in the present context. But his point remains that such roles endure and are deeply important, in creative fields and elsewhere. Likewise, George Morgan and Pariece Nelligan's analysis of creative labour also points towards the existence of *technical creatives* and *creative technicians*, without directly defining this pattern.[7]

Why focus on technical creatives and creative technicians?

These hybrid practitioners require our attention for three reasons. First, observing their existence points to the fact that the creative sectors can make a substantive difference to the quality of local *industrial* production. Likewise, industrial trades, technical knowledge and manufacturing capacity make a significant difference to the quality of *creative* work produced, and to the skill levels of creative practitioners. Technical skills can enable creative ideas to turn into concrete outcomes that are well-executed and ecologically sustainable. The productive and integrated skill relationship between these broadly-termed "creative" and "industrial" arenas demonstrates that these areas need not be seen in opposition, and nor should one area be seen as "historical" (or solely industrial) and the other as "contemporary" (or post-industrial). Rather, creativity and industrial production are, in fact, essential to each other, in ways that Australian culture and governments have thus far failed to appreciate or capitalise upon.[8] In practice, there are myriad crossovers between these supposedly separate worlds – across workshops, studios, factories, vocational training centres and industrial estates – throughout urban, suburban and regional Australia.[9]

Second, a close examination of *technical creatives* and *creative technicians* is significant because it points to the ways in which neoliberalism and deindustrialisation have transformed and fragmented not only job security and working life (this is well-established), but it has also fractured occupational categories and social understandings of skill. Here we find a loose "non-category" of skilled worker, who is not easily placed within formalised occupational sectors or policy settings. This means that these practitioners' precarious occupational experiences are not well understood, and their social, economic and environmental significance is often overlooked.

Third, pointing to the existence of *creative technicians* and *technical creatives* complicates the more mainstream Australian understanding of who a "tradie" is, and what their capabilities might be. In Australian culture, the stereotypical understanding of a "tradesperson" is a younger man – often wearing "hi-vis" – who is typically imagined to be white, heterosexual, able-bodied, manually-skilled, stoic, uncomplicated and who drives a large (petrol) utility vehicle. This cliché is so entrenched in the culture it is regularly leveraged in advertising and political campaigns.[10] But the reality is that skilled practitioners who make, craft, manufacture, repair and maintain are far more complex and diverse than mainstream culture or the media gives them credit for.[11] *Technical creatives* and *creative technicians* and hi-vis wearing "tradies" are not quite the same thing, but there are unacknowledged overlaps here that I seek to surface.

Makers, Manufacturers and Designers *oral history project*

The *Makers, Manufacturers and Designers* (MMD) oral history project – now a part of the National Library of Australia (NLA) collection – featured oral history interviews with craft practitioners, designer-makers and design educators from the states of New South Wales, South Australia and Victoria.[12] Undertaken by myself and co-interviewer Nikki Henningham between 2021 and 2022, MMD traced practitioners' lives and work from roughly the 1960s to the present.[13] A modest number of participants were interviewed: twelve. In Table 5.1, I list participants' roles, occupations, trades and identities to indicate the depth and diversity of experience that one can glean from conversations with just a small number of interviewees. Detailed qualitative "life history" dimensions can provide a more nuanced picture of training, skill, identities and employment decisions.

Oral histories can surface rich contextual understandings, accounting for the complex intersections of class, generation, gender, education, geographic location, cultural and ethnic backgrounds, among other factors.[14] This approach also allows individuals' stories to be situated within broader collective and national histories. As with all oral history material, interpretations of this material should be undertaken with caution, for instance with the understanding that interviews are subjective contexts wherein meaning is co-produced by the interviewer and interviewee in dialogue.[15] In my analysis I am less concerned with whether the interviewees have all their "facts" recalled with precision. Rather, I am more focused on motivations, narrative positioning and modes of understanding work, training, skill and material practice.

TABLE 5.1 Interviewees from the *Makers, Manufacturers and Designers Oral History Project*

Occupations, identities, skillsets and expertise:12 Interviewees from the ***Makers, Manufacturers and Designers*** Oral History Project		
allied health worker	furniture designer-maker	punk
art installer & technician	furniture repairer	retail worker
artisanal woodworker	garment patternmaker	screen printer
artist	garment patternmaking educator	set designer (theatre and film)
assemblage artist	HR recruiter	skate-ramp maker
CNC technician	industrial designer	stone mason (quarryman)
colourist (screen printing)	industrial modelmaker	stop-motion animation technician
community services worker	leatherworker	studio weaver
creative director	mechatronics technician	sustainable fashion activist
defence manufacturing technician	engineering draughtsperson	textile mill worker
design educator	microbiologist	textiles designer
engineering patternmaker	moulding & casting specialist	ticket-writer (signwriter)
fashion designer-maker	nurse	wallpaper factory worker
fibreglass boatbuilder	parent	weaving educator
fine art foundry worker	photography technician (analogue)	youth worker

Precarious work and "doing everything"

The ubiquity of precarious work experiences in neoliberalised economic contexts should not mean we accept this condition *carte blanche*. One core theme emergent from the *MMD* interviews is that *all* those I interviewed had led, and some continue to lead, lives involving insecure work, periods of little income, and balancing short-term contracts so as it make ends meet.[16] Of the participants interviewed for MMD, *all* had experienced (or were continuing to experience) serious precarity, and They each spoke of hard decisions they had to make regarding education and training, housing and chosen occupational sector. A lack of formal employment options – stable, full-time jobs – means that *technical creatives* and *creative technicians* often feel they have no choice but to work as individual contractors and small-business owners, even if they are not entrepreneurially inclined. They are also more likely to take on short-term "jobbing" work for both creative and industrial projects, often at the same time: balancing multiple deadlines. This employment status is often the only way to survive as a practitioner in their field. The much-discussed creative industries' "freelancer dilemma" certainly applies here: always anxiously looking for more work, sometimes burning out from overload, and then experiencing periods of too little work, and therefore too little income.[17] Indeed, some participants were craftspeople who had recently made the hard decision to find (paid) work in another sector entirely, ultimately giving up on the idea of being able to obtain consistent and secure employment in their skilled craft, as the following example explains. Finally, these examples also demonstrate that diverse, cross-sectoral skillsets are not always the product of intentional career decisions, but can also be an outcome of precarity and human desperation for security.

Ana Petidis[18]

Ana Petidis became a studio weaver after a career in the community services and welfare sector. As a second educational path, Petidis undertook a Certificate IV and an Advanced Diploma in Textiles Design and Technology at RMIT, Melbourne. There, she discovered hand weaving, falling in love with the technical *and* creative sides of the craft:

> The second year of night school [was] when we had the weave unit taught to us, and my mind exploded … I just fell. I just fell, and it was – I just had so much wonderment about how this is how you create something! … It was delightful … It's physical work. You have to practice … I was like, I *want* to practice this. This is what I want to do. It felt like all the things, the logical, rational things, the process-oriented things, and then the creativity, the colour … there's quite a bit of maths, and there's angles, and … that logical, rational working alongside with colour and visual processing. You know – how am I seeing colour? How do they interact? How does colour change? How does texture work with this? It got all of my senses involved. And I loved it. It just – my brain was awake, and my body was, too. I was using mind and body in this creative process.[19]

Petidis' language itself demonstrates an alignment of technical knowledge and creativity.

Falling in love with weaving was one thing. Finding paid work was quite another. Petidis' family background is relevant here. During Petidis' childhood, her Greek migrant parents ran a garment-making cut-and-sew production business, initially outworking from their home garage in the 1980s, before briefly opening their own garment factory. As a child, Petidis spent time in the home's workspace, playing with fabrics and threads, undertaking simple tasks like snipping overlocker threads from shoulder pads. Unfortunately, Australia's so-called "Recession we had to have" in 1990 intervened to make the Petidis family's garment factory untenable.[20] But the family business left a lasting impact on Petidis, who enthused: "I love, love production! I love process. I love figuring out the problems in a process-oriented sort of way, and I can only think it must be from that time, from that start."[21] Accordingly, after gaining her Advanced Diploma, Petidis sought practical, hands-on work that was connected to local production. This was a difficult position to be in, in a country where textiles, clothing and footwear (TCF) manufacturing has endured major existential challenges, following Australian economic restructuring from the late 1980s onwards.[22]

Petidis reflected that, for herself and her small cohort of fellow textiles graduates from RMIT, the insecurity that faced them was intimidating:

> We were just like, where do we go to work? Like, this is futile. What are we doing? And someone said, 'Oh, I've got to pick up a contract and pay for my mortgage the next few months,' or 'I'm running out of savings,' … some of us were like, 'I don't know that I can leave my day job to make the full leap because I'm a single woman, I need to think about my superannuation.'[23]

As luck had it, for a few years Petidis found employment as a textiles designer at one of Australia's last woollen mills, Geelong Textiles Australia, in regional Victoria. This role – albeit insecure – placed Petidis between creative textiles design and industrial production, where she "loved that constant need to problem solve":

> I was really … fortunate to use the textile design skills, and understanding that craft skill of hand weaving, being able to design for industry in … upholstery fabrics, and then later we designed a whole range of blankets during the pandemic.[24]

At the mill, Petidis collaborated with sustainable design outfit Seljak Brand. Together, they found a use for the mill's "deadstock" yarn, weaving it into retro wool blankets in outrageous colour combinations, and marketing the blankets directly through social media.[25] For a short time, it was a bustling moment of creativity, sustainable materials and local production working in dynamic fusion, despite the context of Covid-19. In 2022, however, the Geelong Textiles family owners sold the factory and business, which marked the end of Petidis' time at the mill.[26]

Possessing comparatively rare skill as a hand weaver, Petidis' one-off woven works – which she creates in her own time – are now in demand by gallery collections,[27] and she has taught textiles design and weaving at RMIT. Like her work in manufacturing, the tertiary teaching work remains sporadic and insecure. Ultimately, Petidis resolved that weaving and textiles could not reliably be her "day job," it was "totally casualised … unstable." At the time of writing, she works full-time in the community services sector:

> There's a saying … how do you make a fortune in textiles? You start with one … That's the kind of reality … it was precarious … my partner and I have other grand plans for ourselves … and needed that stable income and a salary job where I could have sick pay and annual leave … So I went to my other set of skills, my … community service skills.[28]

The loss of Petidis' technical and creative skills in textiles design and weaving – to both the textiles manufacturing and vocational education sectors – is considerable (to be fair, it is a win for community services).

Claire Tennant

Moulding and casting specialist Claire Tennant likewise reflected on her career and explained the various stages she experienced before she ultimately started her own business (in specialist moulding and casting). Tennant's pathway is perhaps unusual for a moulder. In Australia, moulding is a formal industrial trade, and one that is ordinarily male-dominated. Typical training consists of an apprenticeship (incorporating a vocational Certificate III in Engineering: Casting and Moulding – although nowadays not all foundry moulders hold a formal qualification in the area, largely due to training unavailability).[29] Tennant, by contrast, initially undertook a tertiary degree in industrial design, but decided she did not want to work as a professional designer. In the early stages of her career, she developed local and international "hands-on" experience working in plastics manufacturing, modelmaking for stop-motion animation, and making props for the film industry.

Tennant's experience working at a stop-motion animation outfit in Adelaide was sporadic, as is a great deal of work in the film and television sector:

> It was more about their lack of work than anything I was doing … I was in and out there a few times, but I actually spent the majority of my time working as a receptionist at a kung-fu academy.[30]

Later, Tennant worked at Mackinnon & Saunders in Manchester (one of the UK's best animation producers and puppeteers), which she describes as "my peak career highlight," despite the fact that "the wages were so low for the work that people put in."[31] When her visa expired, Tennant returned to Australia and tried to find work:

> I just couldn't get a job to save my life, and with this glowing CV and all this amazing experience, there just wasn't a job to have. So many people I talked

> to were like, 'If there was a job, you would have it. It would be yours.' But there wasn't, so I just had to get a job. I got a boring job … Then I got a less boring job in an art shop.[32]

Tennant's experience speaks to a lack of opportunities in more than one sector in Australia: both manufacturing/making but also the film and television industry. Finally, Tennant found work in a fine art foundry in Sydney, where she honed her skills in moulding and casting large sculptures, developing expert knowledge of casting, moulding materials and small-run production methods. At this time, Tennant gradually accrued her own contacts and began to take on small jobs independently.

Tennant ultimately decided to take charge of opportunities herself, starting her own business and studio, specialising in moulding, modelmaking and casting. The following quote by Tennant is unusually long, but it is shared because it demonstrates the way in which employment stability for Tennant is paradoxical because requires *doing everything at once*. It also means embracing both technical/industrial jobs as well as more "creative" ones. Tennant's story also points to the difficulty of classifying her field and expertise in a way that acknowledges her broad, cross-sectoral contribution, spanning many fields of practice. Tennant reflected:

> I'm absolutely a jack of all trades. I think my success in running my own business for the last fifteen-odd years has been in that I have so many different … I don't want to say 'expertise,' but yeah, I sit in a very niche spot where there's not really anyone that does so many things and is … the jumping-off point before … factory manufacture. … On the list of things that I do: … mould making for sculpture and casting … people come to me a lot for materials testing. I do life casting, which is taking moulds from people, like baby's hands and feet, and pregnant bellies, and faces, and full head casts for prosthetics for film, but also for body casting, for sculpture that people can hang on their walls … I do slip-casting, slip-cast moulds for ceramicists … I do mould making for bronze and also glass investment casting … foundry casting. I do 3D printing and laser cutting. I teach, as well. I teach at university and also in high schools, and I do art workshops as well, for funsies. I embed things in resin for … display purposes. I do ad work. I do some film work. I have done numerous TV series sets and props over the years … and I also make art for myself, which I've only just started doing in the last couple of years. I never considered myself an artist. More a technician, until recently. Maybe now 30–70 [percent] towards technician (slight laugh) and 30 percent artist … I do a reasonable amount of work for Sydney Vet Science Department, Sydney Uni, for their teaching models. I've done a very complex equine teaching model (slight laugh), which was quite the brain bake, because I had to learn about horse anatomy and all sorts of

> stuff … it's a lot like my brain is full all the time, and there's not a time where I don't have eight jobs on the go … You've just got to compartmentalise a bit, but it's hard![33]

There are evidently high risks of burnout in such an approach. A couple of years after this interview, I was again in contact with Tennant, who explained that this way of working (referred to above) was not sustainable. In the intervening years, it had resulted in serious physical and mental harm. Tennant now operates differently. With her family she has moved out of a high-cost, high-stress urban environment in Sydney, to regional Victoria, and she is more selective about how she works:

> I have since massively cut down on what work I do and who I do it for. I still have a large wheelhouse of skills to offer, but I'm offering them more selectively and only investing in clients that can appreciate the complexity of what they are asking and appreciative of my skills to get it done.[34]

This quote points to the fact that public perceptions of skill *matter* – they matter to workers, and they matter to productivity.

From one practice to another

As noted earlier, the skillsets and knowledge held by *technical creatives* and *creative technicians* are acquired through a diverse mix of formal educational pathways (apprenticeships, degrees) alongside more informal learning pathways, such as auto-didacticism and "on the job" learning. The next two MMD interviewees are practitioners with diverse employment and training histories, who have shifted from one thing to the next in succession.

Mike Buick

Arts technician and assemblage artist Mike Buick formally acquired skills through various vocational training programmes, but is also self-taught in other respects. Buick is a trained composites boat builder, mechanical engineering draughtsperson, analogue and digital photographer and woodworker, and at the time of interview he worked as a casual art installer/technician for a well-known Australian sculptor, while continuing his own art practice. Reflecting on his formative early years, Buick spoke about the value of Australia's vocational educational system (known as TAFE) for his skill development:

> I think it was really useful for me to have that as a start … it was very good. I mean, I had … I'd basically started three careers through TAFE. So, it gave me an amazing opportunity in life … The first career in boatbuilding

was through the Reinforced Plastics Certificate that I did, and the second career in drafting … the engineering drafting course … the third TAFE course I did was the Mechanical Technology Certificate, and that, I think, enabled me to do a lot of work in the art field, as an arts technician, as an installer.[35]

Buick's background draws attention to the fact that Australia's vocational educational system historically operated as a conduit between both industrial and creative skill development.

In the 1990s Buick worked at Gilbarco, a company that manufactured petrol station equipment. He was based in the drawing room as a mechanical draughtsman, hand-drawing pumps, consoles and bowsers, but it was not long until technological change was a factor to contend with:

In my area, definitely … CAD was coming in, but on the factory floor, not very much yet, at that stage … Initially it was just more and more drawings being done on CAD, but there was always hand work to be done.[36]

Economic downturn and company downsizing led Buick into a rocky time, where he was retrenched, rehired, and ultimately decided to pursue other things:

Well, [the job] had two endings, and two beginnings … there was one point where I was retrenched because of a downturn, and I'd spent about three months out of work at that point, looking for work and not being successful, and eventually got the same job back again, because one of the other draughtspeople left, so that was a new beginning in the same role. The other end was when I got the travel bug and decided to go off cycling around the world for a year.[37]

Buick's global cycling tour also led to a passion for photography, which moved well beyond a tourist's enthusiasm for travel snaps:

I'm the kind of person who, when they get interested in something, will launch into it and it will become a passion, so it wasn't long before I was consuming photography magazines and consuming books from the library on photography, and on coming back from that trip I remember regularly going to the library and getting four or five books at a time on photography, technical books, and consuming them and then three weeks later going back and getting another four or five books. That's basically how I learnt photography.[38]

After teaching himself photography (both analogue and digital), this led to a phase where Buick worked as a photography assistant and freelance commercial photographer. As with engineering draughting, Buick's encounter with photography sat him at the cusp of the technological change from analogue to digital. He had to learn both methods, but saw himself as "quite fortunate to have experienced that transition."[39]

Finally, Buick's Mechanical Technology Certificate from TAFE led to contract employment as an arts technician; he now takes on regular casual work an exhibition installer and sculpture production assistant. He also repairs timber furniture on an ad-hoc basis. As Banks observed, arts technicians are one of the most ignored forms of cultural and creative labour,[40] but Buick did not express any concern about lack of recognition. The work is not enough, however, to satisfy Buick's own creative impulses, and he makes sure to only work part-time, so he has enough time to work on his own art practice, making neatly balanced three-dimensional works with reclaimed timber, found objects and two-dimensional collage. Buick explained:

> A lot of the artwork that I do now harks back to mid-century design on the one hand, but on the other hand, [it] can have a bit of an engineering bent because of my background and my experience. You're a product of your experience and your past, you know.[41]

Clearly, Buick is well aware of how his technical and industrial experience informs his creative practice.

Karl Bergersen

Karl Bergersen, proprietor of the Sydney-based recycled timber furniture business Three of a Kind, has a life history that traverses multiple subcultures, creative practices, manual trades and industries. In many ways, his career evidences a wild mixing of late twentieth-century Australian and American subcultural and production histories. Throughout his life, Karl's interests, fields of expertise and making experiences include: skateboard ramp design, screen-printing, leatherwork, stone masonry, furniture design, woodwork, graphic arts and music (band) management. But to even write such a list feels reductive, as much has been left out. Bergersen was one of the few makers I interviewed who has shifted from a precarious individual maker – scraping by on very little – towards a successful small business owner with a number of employees. He laughingly describes himself as a "cashed-up bogan,"[42] self-consciously referring to the ways in which family background, taste and place all influence social class categories. In this way, Bergersen himself is well aware that he is hard to categorise, and appears to enjoy an element of surprise when encountering people.

Bergersen proudly explained that, while he has some official vocational training in screen printing and stonemasonry, his current career as a recycled furniture designer and manufacturer emerged from proudly self-taught roots. Here, "ticket" refers to formal vocational qualifications:

> I've got two tickets, but I don't use either of them. I've got my Quarryman's ticket … I don't use that … So I've got two tickets in two different trades that I don't use, and I've got a furniture store with no ticket, and that's what I do.[43]

Unlike Buick, Bergersen is less enthusiastic about the training provided by Australia's vocational training institutes, nor about formal education generally. This position also impacts Bergersen's views about hiring labour for his furniture-making business:

> It's hard to get the right people. I find it's easier, rather than getting qualified cabinetmakers and joiners and teaching them to un-learn, and adopt my aesthetic, I find it easier to get people like builders, or [basic] carpenters … and then teach them how to – because they don't have a preconceived idea … The people who I've found to be the best at this kind of work are … set-carpenters. People that have worked in film and TV. …They're … taught to work to the aesthetic.[44]

Here is a direct example where a local designer-manufacturer has specifically sought *creative technicians* (set designers/set-makers), due to their aesthetic understanding.

For Bergersen, employment and subcultural involvement have always been intertwined, and I draw attention to this to indicate the *cultural* complexity of Australian local production between the 1970s and 1990s. As a youth, Bergersen was involved in surf culture and early skate culture in Sydney in the 1970s. He worked as a "general dogsbody" at Emerald Surfboards, and discussed early experiences building DIY skateboard ramps, and skating vertical at Pymble Pools with skateboarders Adrian Jones and Johnny Tesoriero. Surfing and skating went together, but Bergersen also embraced "gear head" car culture, and soon after that, punk. Later, Bergersen's interests expanded to include Santa-Fe styling, Rodeo and cowboy culture, and he is a passionate collector of mid-century Americana. This unique mix of subcultural interests and passions is something Bergersen consciously articulated, noting that his wide-ranging passions did not always fit smoothly in Australian culture:

> I've always mixed up my passions irrespective of the fact that the culture, or the tribes and the people didn't mesh so well. They do now, but they didn't back then. And that's why punk rock was so appealing and serendipitous to me: because there was no fucking rules. It was, do whatever you want … and that appealed to me a lot, you know. That is just in my nature, I think. So,

> punk rock, yes, it was a movement, but I think I've always been a punk rocker, even before it existed.[45]

Bergersen's early employment history also included several years working at Sydney t-shirt factory Phantom Textile Printers, where he excelled in the technical sides of colour mixing, as well as being a "prankster, a clown and a buffoon" (his self-description).[46] Consequently, he was fired and re-hired a number of times by proprietor Dare Jennings, better known (later) as the founder of Australian brands Mambo and Deus ex Machina. Bergersen's work history also involved working for local councils building stone retaining walls around Sydney Harbour. Later he became a self-employed leatherworker, sitting on his front porch handmaking belts for the Zimmerman sisters' local fashion shop, before they had hit major fashion success. The time spent making belts was pleasurable but precarious: "I was on the, I wouldn't say 'bones of my arse' but I was living – sailing pretty close to the wind in a financial sense."[47]

Bergersen's foray into recycled furniture began in the leatherwork period, when he needed a table for himself, and, having little means, decided to make one. He "borrowed" some timber waste from a construction site across the road, and produced a solid, well-proportioned recycled timber table with its own unique character. Soon, Bergersen's distinctive recycled furniture style was attracting attention, and he began making pieces for friends and acquaintances. Ultimately, Bergersen decided to start his own furniture business and opened a shop, initially in South Newtown, Sydney. It was not an easy start:

> I've never worked harder in all my life … It was fucken hard … It's the first time I had my own business … and I was thinking about my future. Am I making a go of it, or not? There's your motivation right there, you know? Like, I couldn't afford to take my foot off the gas because I had overheads … When I opened that shop, I had no money … I had a tax return of 500 bucks.[48]

The hard work ultimately paid off, over time. Bergersen's furniture business is now at a stage where he comfortably refuses to take on new work when the backlog is too long, because he does not want his customers waiting for long periods.

Like the craft entrepreneurs described by Susan Luckman,[49] Bergersen consciously resists expanding his business any further, not wishing to lose creative control: "You've got to get a balance where…all this furniture has me in it, my stamp on it and my aesthetic and my input, and once I get too far from the factory floor, that's when things go south."[50] In this way, Three of a Kind stays relatively small in scale. One of Bergersen's bugbears is that in the current globalised economy – where cheap, poor-quality furniture can be ordered from overseas and arrive in a flatpack in a matter of days – people no longer understand the true costs of carefully crafted, local production. By contrast,

Three of a Kind is an all-in-one set-up with the shopfront, workshop and Bergersen's home upstairs, now based along the bustling Parramatta Road, in Leichhardt, Sydney:

> Retail at the front, and then workshop at the rear with the glass at the back ... which has been a huge benefit, just really driving home the point that it's made on the premises. People are a bit taken aback. It's really rare to see anyone manufacturing in Australia, and it's also really rare to come into a premises and you're looking at, and talking to, the person that's making the stuff for you.[51]

This set-up allows him to *perform* production, as well as emphasise the technical nature of how recycled timber furniture is made.

For all his talk of hard partying days, Bergersen is no loose cannon in terms of craft. He has a strict aesthetic sensibility, strong craft ethic and longstanding tacit technical knowledge, which he is cautious about imparting to others unless he really trusts them. Bergersen also knows where to source good quality recycled timber in Sydney (it is all about personal connections), and how to cut, treat and paint recycled timber so that it appears consistently aged, thus retaining its character. With this aesthetic sensibility comes a strong commitment to making durable furniture that should endure over generations: "I want stuff out there ... that is a lifetime purchase. Planned obsolescence just isn't in our vocabulary. Pieces that people will be proud – heirloom stuff."[52] Bergersen proudly boasts that he has now been in the furniture business long enough that the children of his early buyers return, as adults, to report on how the family piece is going.

While this isolated story is one of success, it is also rare: most furniture-making and furniture consumption does not look like this, and Bergersen's pieces face that long-standing "Arts and Crafts dilemma": most pieces are only affordable for reasonably wealthy buyers, as their pricing accurately reflects the labour involved in their production. Bergersen, for his part, wishes that mainstream Australian culture understood and respected craft skills – and the intense efforts involved in dealing with recycled timber – to a greater degree. Despite possessing abundant practical and technical proficiencies, *creative technicians* and *technical creatives* remain paradoxically undervalued, even as demand for their expertise grows. Such practitioners exist in a neoliberalised political-economic system that does not meaningfully account for their skills or capacities.

Conclusion

As noted at the outset, the example of *creative technicians* and *technical creatives* is indicative of larger problems in how we understand, conceptualise and plan for the changes to working life and skill in the future. The intertwined challenges of climate change, geopolitical and economic instability and technological shifts combine to expose just how much our existing categories and conventions of occupation, skill and education are no longer fit for purpose. This begs the question: where to from here? One place to start is from the ground up: what capacities do people actually have, when we actually take the time to observe and listen?

Building an understanding of *technical creatives* and *creative technicians'* life histories and personal perspectives helps open our awareness – beyond just the creative industries – to people's ability to develop multiple sets of practical skills over a lifetime. Much like the "maintainers" and repairers increasingly studied by scholars in a variety of disciplines,[53] *technical creatives* and *creative technicians* possess manual and process skills, such as sustainable making practices and knowledge of older technologies, infrastructures and production systems, as well as social and relational knowledge of supply networks and sustainable waste management chains.[54] Such dispositions will be critical in the context of futures featuring infrastructural breakdown, disaster recovery, adaptation and finite resources.[55] One of the key things that the lives of *technical creatives* and *creative technicians* tell us is how they bring their prior experience to bear on whatever they undertake next. Ignoring such capacities would be a great loss, because the skills that such practitioners bring to society are not isolated to one profession or industry, but are transferrable in ways that will likely be valuable for years to come.

Notes

1 Stein and Carr, "Call for Papers." In using "creative" in this chapter, I acknowledge that the term has, for several decades, become a buzzword within the politics of neoliberal economic restructuring. "Creative" therefore carries the stain of political rhetoric, false hope about opportunities in creative economies, and the supposed benefits of flexible labour. I continue to use the word, however, because it is useful in collectively encompassing the arts, craft, design and aspects of media and communications. For critical commentaries on the neoliberal mobilisation of "creative" and "creativity," see, for example: McRobbie, *Be Creative*; Morgan and Nelligan, *The Creativity Hoax*; Gill & Pratt "In the Social Factory?"; Hawkins, *Creativity*; Taylor & Luckman (eds), *Pathways into Creative Working Lives*; Banks, Gill and Taylor (eds), *Theorizing Cultural Work*; among others.
2 Carr and Gibson, "Geographies of Making."
3 Luckman and Tower, "The Value of Craft Skills to the Future of Making in Australia."
4 The background concepts that inform this chapter have developed from my empirical research over the past decade and a half, which has largely focused on the human and design outcomes of deindustrialisation and economic neoliberalisation in Australia from the 1980s onwards.
5 I elaborate on this particular pathway in Stein, *Industrial Craft in Australia*, 191–224.
6 Banks, "Craft Labour and Creative Industries."

7 Morgan and Nelligan, *The Creativity Hoax*.
8 Luckman, "Making Bodies"; Carr and Gibson, "Geographies of Making."
9 See also: Grodach, O'Connor, and Gibson, "Manufacturing and Cultural Production,"; Gibson et al. "Made in Marrickville."
10 Humphrys, Stein and Frankham, "The Deep Political Power of Fluoro."
11 Luckman, "Making Bodies."
12 Stein and Henningham, *Makers, Manufacturers and Designers Oral History Project* (hereafter: MMD), National Library of Australia. Part of an Australian Research Council Discovery Early Career Research Award (ARC DECRA 2021: DE210100158), which uncovered connections between the design industry, technical trades and vocational education, tracing the second half of the twentieth century to the present.
13 In line with the preferred style used by the NLA, *MMD* uses life-history interviews, the longest in this project being over five hours and the shortest at around two and a half hours' duration.
14 The nature of life history oral histories is that, in many cases, participants choose to remain identified. In limited cases I occasionally anonymise participants, where ethical concerns may arise.
15 For a nuanced (and still relevant) summary of different approaches, sub-disciplines and evolving concerns in oral history practice over the past five decades, see Thomson, "Four Paradigm Transformations in Oral History."
16 These findings are certainly in line with a great deal of existing research into cultural labour from the 1990s onwards, see, for example: Gill and Pratt, "In the Social Factory?"
17 Pitts, "Creative Labour, Before and After 'Going Freelance.'"
18 Another aspect of Petidis' story is shared in our sister collection, see: Stein, "Saving the Loom."
19 Petidis, interview with Henningham, *MMD*.
20 Petidis, "Devoted."
21 Petidis, interview with Henningham, *MMD*.
22 Webber and Weller, *Refashioning the Rag Trade*.
23 Petidis, interview with Henningham, *MMD*.
24 Stein and Petidis, "Rescuing Weaving Technologies."
25 Seljak Brand, "This Geelong Mill Has Been Weaving Wool Fabric for 100 Years"; Seljak Brand, "Meet the Makers."
26 "Geelong Textile Group Sold to ATI."
27 For example, Art Gallery of Ballarat acquisition, Petidis, *Devoted*, 2021, https://www.artgalleryofballarat.com.au/explore/the-collection/search/12636.
28 Stein and Petidis, "Rescuing Weaving Technologies."
29 Romensky, "Hard, Dirty Foundry Work Copes with Digital Disruption and Lack of Apprenticeship Courses."
30 Tennant, interview with author, *MMD*.
31 Tennant, interview with author, *MMD*.
32 Tennant, interview with author, *MMD*.
33 Tennant, interview with author, *MMD*.
34 Tennant, personal communication with author, 28 November 2024.
35 Buick, interview with author, *MMD*.
36 Buick, interview with author, *MMD*.
37 Buick, interview with author, *MMD*.
38 Buick, interview with author, *MMD*.
39 Buick, interview with author, *MMD*.
40 Banks, "Craft Labour and Creative Industries."
41 Banks, "Craft Labour and Creative Industries."

42 Bergersen, interview with author, *MMD*. "Bogan" is an Australian slang term, loosely referencing unsophisticated working-class culture; it can be pejorative, humorous, self-depreciating or celebratory, depending on usage.
43 Bergersen, interview with author, *MMD*. A quarryman, in this usage, refers not to someone who works in a quarry, but a lower level of qualified stonemason. It is a trade related to cutting and placing stone for retaining walls and pavers.
44 Bergersen, interview with author, *MMD*.
45 Bergersen, interview with author, *MMD*.
46 Bergersen, interview with author, *MMD*.
47 Bergersen, interview with author, *MMD*.
48 Bergersen, interview with author, *MMD*.
49 Luckman, "Craft Entrepreneurialism and Sustainable Scale."
50 Bergersen, interview with author, *MMD*.
51 Bergersen, interview with author, *MMD*.
52 Bergersen, interview with author, *MMD*.
53 Strebel et al. (eds), *Repair Work Ethnographies*; Russell and Vinsel, "After Innovation, Turn to Maintenance."
54 Carr, "Maintenance and Repair Beyond the Perimeter of the Plant"; Carr, "Repair and Care"; Stokes and De Coss-Corzo, "Doing the Work."
55 Carr, "Repair and Care."

Bibliography

Banks, Mark. "Craft Labour and Creative Industries." *International Journal of Cultural Policy* 16, no. 3 (2010): 37–41.

Banks, Mark, Rosalind Gill and Stephanie Taylor. *Theorizing Cultural Work: Labour, Continuity and Change in the Cultural and Creative Industries*. London: Routledge, 2013.

Carr, Chantel. "Maintenance and Repair beyond the Perimeter of the Plant: Linking Industrial Labour and the Home." *Transactions of the Institute of British Geographers* 42, no. 4 (2017): 642–54.

Carr, Chantel. "Repair and Care: Locating the Work of Climate Crisis." *Dialogues in Human Geography* 13, no. 2 (2023): 221–39.

Carr, Chantel and Chris Gibson. "Geographies of Making: Rethinking Materials and Skills for Volatile Futures." *Progress in Human Geography* 40, no. 3 (2015): 1–19.

"Geelong Textile Group Sold to ATI." *Ragtrader*, 20 June 2022. www.ragtrader.com.au/news/geelong-textile-group-sold-to-ati#:~:text=Geelong%20Textiles%20and%20Geelong%20Dyeing,West%20of%20New%20South%20Wales

Gibson, Chris, Carl Grodach, Craig Lyons, Alexandra Crosby and Chris Brennan-Horley. "Made in Marrickville." Urban Cultural Policy and the Changing Dynamics of Cultural Production (ARC Discovery Project Report), QUT, University of Wollongong and Monash University, 2017. https://documents.uow.edu.au/content/groups/public/@web/@socs/@dgsc/documents/doc/uow270878.pdf

Gill, Rosalind and Andy Pratt. "In the Social Factory? Immaterial Labour, Precariousness and Cultural Work." *Theory, Culture & Society* 25, no. 7–8 (2008): 1–30.

Grodach, Carl, Justin O'Connor and Chris Gibson. "Manufacturing and Cultural Production: Towards a Progressive Policy Agenda for the Cultural Economy." *City, Culture and Society* 10 (2017): 17–25.

Hawkins, Harriet. *Creativity*. London and New York: Routledge, 2017.
Humphrys, Elizabeth, Jesse Adams Stein and Bettina Frankham, "The Deep Political Power of Fluoro: How Hi-Vis Became a Symbol of Working Class Masculinity." *The Conversation*, 18 September 2024. https://theconversation.com/the-deep-political-power-of-fluoro-how-hi-vis-became-a-symbol-of-working-class-masculinity-238584
Luckman, Susan. "Craft Entrepreneurialism and Sustainable Scale: Resistance to and Disavowal of the Creative Industries as Champions of Capitalist Growth." *Cultural Trends* 27, no. 5 (2018): 313–26.
Luckman, Susan. "Making Bodies, Craft Skills and the Legacies of Policy 'Blokeism.'" *Continuum* 37, no. 5 (2023): 595–607.
Luckman, Susan and Ash Tower. "The Value of Craft Skills to the Future of Making in Australia." *Report. Creative People, Products and Places Research Centre (CP3)*, University of South Australia, 2023. https://apo.org.au/node/324171
McRobbie, Angela. *Be Creative: Making a Living in the New Culture Industries*. Cambridge & Malden, MA: Polity Press, 2016.
Morgan, George and Pariece Nelligan. *The Creativity Hoax: Precarious Work and the Gig Economy*. New York and London: Anthem Press, 2018.
Petidis, Ana. "Devoted: Adding Weight to Weaving." *Garland*, 2 September 2022. https://garlandmag.com/article/devoted-adding-weight-to-weaving/
Pitts, Frederick Harry. "Creative Labour, Before and After 'Going Freelance': Contextual Factors and Coalition-Building Practices." In *The New Normal of Working Lives*, edited by Susan Luckman and Stephanie Taylor, 87–107. Cham: Springer, 2017.
Romensky, Larissa. "Hard, Dirty Foundry Work Copes with Digital Disruption and Lack of Apprenticeship Courses." *ABC News*, 30 September 2018. www.abc.net.au/news/2018-09-30/hard-dirty-foundry-work-coping-with-digital-disruption/10303254
Russell, Andrew L. and Lee Vinsel. "After Innovation, Turn to Maintenance." *Technology and Culture* 59, no. 1 (2018): 1–25.
Seljak Brand. "Meet the Makers at the Geelong Mill We Work with." *Seljak*, 23 September 2021a. www.seljakbrand.com.au/blogs/news/meet-the-makers-profiles-of-the-geelong-mill-workers
Seljak Brand. "This Geelong Mill Has Been Weaving Wool Fabric for 100 Years – Now They Make Seljak Brand Blankets Too!" *Seljak*, 15 July 2021b. www.seljakbrand.com.au/blogs/news/this-geelong-mill-has-been-weaving-wool-fabric-for-100-years-now-they-make-seljak-brand-blankets-too
Stein, Jesse Adams. *Industrial Craft in Australia*. Cham: Palgrave Macmillan, 2021.
Stein, Jesse Adams. "Saving the Loom: Tracing One Machine's 20-year Journey from Strategic Government Investment to Small-scale Craft Volunteerism." In *Designing through Planetary Breakdown*, edited by Jesse Adams Stein and Chantel Carr. London: Routledge, 2025.
Stein, Jesse Adams and Chantel Carr. "Call for Papers." *All Hands on Deck Symposium 2023*, 2022. https://allhandsondeck2023.com/call-for-papers/
Stein, Jesse Adams and Nikki Henningham. "Makers, Manufacturers and Designers Oral History Project (MMD)." Oral History Interviews. Canberra: National Library of Australia, 2021–2022. Catalogue: https://catalogue.nla.gov.au/catalog/8646329. (Interviews with Ana Petidis, Claire Tennant, Mike Buick and Karl Bergersen [all recorded 2022].)

Stein, Jesse Adams and Ana Petidis. "Rescuing Weaving Technologies: Craft and Community Amid Closures and Cutbacks." Conference presentation. *All Hands on Deck*. Sydney, 21 July 2023.

Stokes, Kathleen and Alejandro De Coss-Corzo. "Doing the Work: Locating Labour in Infrastructural Geography." *Progress in Human Geography* 47, no. 3 (2023): 427–46.

Strebel, Ignaz, Alain Bovet and Philippe Sormani, eds. *Repair Work Ethnographies*. Singapore: Springer Singapore, 2019.

Taylor, Stephanie and Susan Luckman, eds. *Pathways into Creative Working Lives*. Cham: Palgrave Macmillan, 2020.

Thomson, Alistair. "Four Paradigm Transformations in Oral History." *The Oral History Review* 34, no. 1 (2007): 49–70.

Webber, Michael and Sally Weller. *Refashioning the Rag Trade: Internationalising Australia's Textiles, Clothing and Footwear Industry*. Sydney: University of NSW Press, 2001.

PART II

Industrial Transformation

6

A TRANSITION WITH TEETH

Reconfiguring the Concept of Just Transition to Recognise the Imperative of Fossil Fuel Cessation

Frances Flanagan

Introduction

To discuss questions of work, class and climate change carries with it the risk of being accused of celebrating the wrong kind of solidarity. In a review of Matthew Huber's 2022 book *Climate Change as Class War*, Michael Levien described Huber's argument that unionised energy workers should be understood as key historical agents in climate change politics as a call that "white energy workers of the north unite!"; an expression of an outdated, crass and teleological Marxism, ignoring Third World and Black Marxism, neo-Gramscianism and Marxist feminism.[1] In reply, Huber denounced Levien's critiques as evidence of an "all-too-typical" detachment from the extent to which decarbonisation involved industrial work and industrial skills that, properly managed, would facilitate a massive expansion of industrial jobs, income growth and unionisation for otherwise abandoned communities.[2] Levien's failure to grasp the potential of existing industrial workers – electricity utility workers, in particular – to organise at the point of energy production and act as the vanguard of the climate proletariat was evidence of "academic *faux*-Marxism" and detachment from the kind of majoritarian climate politics that has the potential to genuinely deliver for poor and marginalised communities. In a world awash with "professional class environmentalism," that patronisingly turns to climate science to "shame" people into action, Huber argued that it was crucial to build a materialist, socialist environmental politics with industrial workers at the frontier. "It is clear to me Levien has little interest in such a political project," Huber concluded, "but I'm sure he will continue to gain accolades among his colleagues."[3]

DOI: 10.4324/9781003512240-9

In such an acrimonious landscape, it is tempting to look on the idea of "just transition" as a kind of safe haven for those seeking to build solidarity between workers and environmentalists. The concept appears to bundle together principles that are virtually impossible for anyone who is not a committed neoliberal to disagree with. Just transition affirms the importance of ensuring that the process of "greening" economies is well managed and contributes to the goals of decent work for all, social inclusion and the eradication of poverty. It insists on participation as a pre-eminent virtue and recognises that frameworks of participation need to extend across multiple scales and aspects of policy making (macroeconomic, industrial, sectoral and labour). It is flexible and adaptive, recognising that "no one size fits all" and that the mix of policies and approaches must be tailored to individual, local contexts.[4] Just transition is also a well-known and popular concept, endorsed by the International Labour Organization (ILO), Paris Agreement and over fifty nations at COP24. On its face, just transition appears to be a solid foundation for reconfiguring systems of work in response to the climate crisis: supple enough to adapt to local contexts, consistent with the environmentally-oriented militancy prized by Huber, and wide enough to encompass the demands for environmental justice articulated by Indigenous and frontline communities, as observed by Levien.[5]

For all its inclusionary promise, though, just transition lacks critical purchase on the fundamental cause of climate change: the fact that the global process of "greening" economies, to date, has substantively amounted to little more than the *addition* of renewable energy sources to existing fossil fuel forms, rather than their displacement. Just transition was not developed to deal with the problem of climate change; it was founded on the more general ideal of "sustainable development" and (in its 2015 ILO Guideline formulation) does not prioritise or even mention fossil fuel extraction as a pre-eminent dynamic of the crisis. Consequently, the term is open to being invoked to describe transitions that potentially perpetuate ongoing fossil fuel extraction. For just transition to "have teeth," it is crucial that it be reconfigured to ensure that it supports genuine energy *transitions*, rather than a mere energy *additions*.[6]

This chapter explains why the just transition concept we have is configured as it is and explains the stakes of that configuration. In this chapter, the current ILO just transition principles are argued to be problematic – for actors who are dealing with climate change-motivated work changes – because they: (a) frame climate change as a generic environmental challenge rather than an existential threat to the web of life; (b) emphasise collaboration over conflict and shared responsibility across all employment relations actors; and (c) accept the institutions and categories of labour law inherited from past industrial struggles. After outlining each of these three dimensions, the chapter concludes with a practical proposal for adding an additional, and more powerful objective to received just transition principles, which involves amending the definition of just transition to mean a transition where fossil fuel extraction is phased out within a prescribed time frame, with longer period of time allocated for non-OECD countries.

Just transition's framing of climate change: Just another environmental challenge or an existential threat to life?

For a concept so widely discussed as being crucial to addressing climate change, the 2015 ILO Guidelines – widely seen as the anchor of just transition policies – accord little attention to climate change. Far from being represented as a general destabilising dynamic that threatens the fabric of life, the words "climate change" rather appear in a piecemeal fashion as a potential source of unemployment, an occupational health and safety risk, an impetus to adaptation and a stimulus to resilience policies. While the Guidelines include many injunctions to develop, incentivise, facilitate, encourage and enable green industries, there is no imperative on parties to withdraw support for fossil fuel extraction. Indeed, fossil fuels are not mentioned in the Guidelines at all. It is not climate change, but rather the imperative to pursue "sustainable development" that forms the overarching motivation to just transition principles. That is, development where the economic, social and environmental dimensions of development are treated as being of "equal importance and must be addressed together."[7] Subsequent academic analyses of just transition principles have tended to reinforce the tendency within the Guidelines to treat climate change as a generic environmental problem, rather than one with overarching and distinctive characteristics. Scholars have, for instance, evaluated different just transition case studies in terms of generic measures such as "boldness," "timeliness," "impact" and "measurability."[8] Dimitris Stevis has suggested that just transitions principles are broadly relevant to any kind of industrial transition, including transitions that are wholly non-environmental in motivation, such as the school, supermarket or automobile plant.[9]

In many ways, just transition's non-engagement with the distinctive contours of climate change as an overarching environmental paradigm is unsurprising, given that it developed as a response to non-climate change environmental problems. The initial concept of a "Superfund for Workers," later renamed a "just transition," emerged from US Oil, Chemical and Atomic Workers Union (OCAWU) Secretary Tony Mazzocchi's argument in the early 1990s that that law should accord recognition and remedial funding to the families of workers left behind by environmental initiatives to shut down toxic pesticide plants, just as it offered resources for the remediation of contaminated landscapes.[10] Were state provision made to such workers for retraining, relocation, reskilling and redundancy, the "jobs versus environment" binary would be revealed to be illusory. When the term was taken up on a global scale in a climate change context, by the International Confederation of Free Trade

Unions (ICTFU) at COP3 in Kyoto 1997, it followed a similar logic to OCAWU's claim following the shutdown of individual toxic pesticide plants: a reactive and remedial demand in response to environmental policies already implemented directed towards limiting the economic and social costs of those policies on working lives.[11]

The emphasis on the generic environmental dimensions of climate change in just transition is at odds, however, with the imperatives that have been articulated by climate scientists.[12] The precise and prescriptive contours of fossil fuel phase down as the *sine qua non* of sustainable development post-dated the development of just transition by some two decades.[13] While there have long been descriptions of "remaining carbon budgets" for limiting warming to 1.5 degrees,[14] a translation of those budgets into explicit positive imperatives on states – do *not* grant licences; do *not* subsidise extraction – have been more recent. In 2021, the International Energy Agency stated that no new unbated coal, oil or gas developments could be approved if the world is to limit temperature rises to 1.5C above pre-industrial levels.[15] In 2023, the UN Secretary-General added to the IEA's imprimatur for no new coal projects with a call for phasing out existing projects within a particular timeline. Coal should be phased out by 2030 in OECD countries and 2040 in all other countries. All international public and private funding of coal and licensing or funding of new oil or gas should cease. There should be an end to the expansion of existing oil and gas reserves, a shifting of subsidies from fossil fuels to a just energy transition, and a global phase-down of oil and gas production compatible with the 2050 global net zero target.[16]

Informing all of these conclusions is an awareness that, regardless of the introduction of renewable energy sources, the longer fossil fuels continue to be extracted and combusted, the higher the probability that climate system "tipping points" will be crossed, and the planet will enter a "hothouse Earth" scenario. In this scenario, climate stabilisation becomes beyond the control of humans.[17] To date, fossil fuel phase-out commitments have not been made by key extractive states such as Indonesia, Australia, India, the United States and China. These states continue to grant licences for new projects[18] and publicly subsidise extraction; global fossil fuel subsidies were estimated to be $5.7 trillion in 2020 (6.8 percent of global GDP) according to the IMF in 2021.[19]

Against this context, just transitions language, with its non-prescriptive objective of "greening" economies, has been used to legitimate continued fossil fuel extraction. Major fossil fuel firms and lobby groups such as the World Coal Association have claimed to support a "just energy transition" through its endorsement of "efficient" coal combustion and its advocacy of the role of coal in poverty alleviation.[20] Woodside, a major Australian oil and gas operator with comprehensive plans for new offshore extractive gas projects, describes itself as contributing to just transition by virtue of its contributions to "providing decent work," "stakeholder engagement" and the provision of "lower carbon and affordable energy."[21] The term has also been used by unions with a high proportion of fossil fuel extractive workers as a basis for arguing that the transition away from fossil fuel extraction must be slow. Officials of the IG Bergbau, Chemie, Energie (IG BCE) union in Germany, for instance, were found by researchers to argue that a "just" transition was "a slow transition that preserves the status quo as long as possible."[22] Similarly, the (then-called) Australian Construction Forestry Maritime, Mining and Energy Union (CFMMEU) have been conspicuous supporters of the just transition concept,[23] but have selectively defined it to apply only to the phasing out of brown coal extraction for domestic electricity conception. Meanwhile, they simultaneously insist that metallurgical coal for export is an industry with a long-term future that will be "in demand for decades to come," including because of the importance of steel for renewables infrastructure.[24] Such arguments do not engage with the implications of open-ended, market-led coal extraction in terms of locking-in carbon intensive steel production supply chains globally and domestically,[25] its disincentivising effects on the development of green steel technologies,[26] as well as the extent to which many mines designated as producing metallurgical coal also produce large volumes of thermal coal.[27]

Just transition emphasises collaboration over conflict

Just transition has thus become a framework for not only debating the terms by which workers may be supported after the imposition of climate change policies that wind down extractive industries, but also for debating *whether* and *when* fossil fuel industries should wind down *per se*. Just transition also frames climate change-motivated decision-making as ideally consensual and collaborative, rather than conflictual. Such negotiations are conceptualised as encounters where all parties share a common capacity to spontaneously "choose" just transition pathways, regardless of the particular time, place, context and the legal obligations and constitution of each entity. These implications are reinforced by narratives that are commonly rehearsed about paradigmatic "good" transitions that involve long-term tripartite workforce planning and pro-environmental union behaviour. The example of hard coal mine closures in the Ruhr Valley, for instance, where long-term coordinated planning meant that forced redundancies were wholly avoided, is often referred to as exemplifying the idea that just transition is possible given sufficient political will.[28] The case of Spain is similarly commonly cited, where an eight-year *Plan del Carbón* was concluded with EU support over five coal mining regions (involving comprehensive provision for early retirements, redundancy payments and environmental regeneration work in former mining sites).[29] In narrating such case studies, the significance of non-environmental motivations for coal mine closures tends to be understated. Hard coal mining in the Ruhr primarily ceased as the result of declining economic competitiveness[30]; as were coal mine closures in the Upper Silesia region of Poland[31] and in Spain.[32]

Similarly, the extent to which ostensibly "good" transitions actually involved continued fossil fuel extraction, or a shift from one kind of fossil fuel to another, is generally minimised. In the Netherlands Limberg region, for instance, the closure of coal mines was consensually accepted *because* of widespread acceptance that another fossil fuel energy source, gas, would replace it.[33] Most Hazelwood power station employees in the Latrobe Valley, in the Australian state of Victoria, were transitioned into other kinds of fossil fuel employment.[34] These narratives also tend to place limited emphasis on the ambivalent role played by actors in phasing out industry. In the case of the Ruhr, for example, States and unions made extensive efforts to delay cessation and preserve coal and steel production (including through the payment of extensive subsidies).[35]

The notion that a collective, consensus-based decision to phase out extraction on orderly and planned terms is readily available as a policy choice for incumbent employers, states and unions also overlooks the asymmetries of power and responsibility concerning fossil fuel extractive projects and the labour such projects require. Decisions about whether or not to open new fossil fuel resources or extend existing projects are overwhelmingly made by state actors, who are responsible for making, taming and shaping energy markets.[36] It is States who largely own fossil fuel reserves and have the right to issue and terminate mining leases. State actors are the ultimate regulators of industrial relations and energy systems. They are the primary owners, globally, of extractive infrastructure.[37] They make decisions over the payment and terms of fossil fuel subsidies and have the power to enter international trade agreements that shape the economic and legal context for extraction. States also determine the terms of environmental law and the terms by which fossil fuel mines are granted conditional permission to operate.

By contrast, many fossil fuel workers are members of unions that are, by their constitutions, existentially entwined with the fossil fuel industry. The current objects of the main union representing coal miners responsible for coal exports in Australia, for instance, the Queensland Mining and Energy Union (MEU), requires the union to "endeavour, by all lawful means … to carry on, or participate directly or indirectly and alone or with others the carrying on of any mining or prospecting operation and related activities thereto" and to "watch over, improve, foster and protect the industrial and social interests of the membership," who are defined as employees (inter alia) of the coal, shale, oil and gas industries.[38] While it is common in just transition discourse to point to past instances of environmentally-protective union action (such as the Australian Builders Labourers' Federation's refusal to work on environmentally destructive urban projects,[39] IG Metall's participation in environmental debates emerging from the 1971 Stockholm Conference, or the United

Automobile Workers support for Earth Day in 1970[40]) recognition of the legal constraints on unions are less observed. These powerful legal limits are not diminished by the fact that many occupations in fossil fuel extraction involve skills and aptitudes that afford many workers the theoretical possibility of employment transition into non-fossil fuel extractive contexts.[41] Fossil fuel employers are similarly constrained in their ability to collectively phase down fossil fuel extraction on a categorical basis as a result of the operation of Corporations Law, Mining Law, contract, fiduciary obligations, international trade law as well as material interests.[42]

Just transition does not attend to the legacies of fossil fuel-based capitalism encoded in existing labour law

Every potential fossil fuel ore body that must stay in the ground is subject to laws that shape the outlook of extractive workforces. These laws have been overwhelmingly inherited from the fossil fuel era.[43] To the extent that existing just transition discourse invites a fantasy of spontaneous consent to fossil fuel cessation, it fails to recognise that the categories and institutions of labour law are not a neutral and unchanging "backdrop" to transition, but rather constitutive elements of carbon-intensive industrial systems that demand, in themselves, debate and reconfiguration in light of the climate crisis.

Nearly every element of labour law, including its recognition of unions and the rights accorded to them, can be said to have been shaped by a set of socio-ecological paradigms that associated fossil fuels with the stable reproduction of life and narratives of progress and social prosperity.[44] The imprint of laws and institutions powerfully shape the actions that may be feasibly available to actors, the potential flows of solidarity between different groups of workers and others, and the material stakes of supporting particular climate policies.

Australia is a liberal market economy with an uncoordinated labour market, where there is no lawful basis for unions to engage in industrial action in support of workers in another sector, nor any current capacity for wage gains in one sector to automatically "flow" to another.[45] Fossil fuel extraction (and coal extraction, in particular) was entwined with the original recognition of labour rights that Australian colonial systems inherited from the United Kingdom,[46] and fossil fuel extraction-related occupations were specifically privileged through the use of public power to support essential industry. During the Second World War, Australian Prime Minister John Curtin established coal, maritime and stevedoring commissions to ensure secure wartime production.[47] Postwar governments drew a nexus between coal mining, industrialisation and the aspiration to achieve a larger national population.[48] Coal production had a pre-eminent place in the tariff protected manufacturing-based nation-building projects of the mid-twentieth century, which saw the introduction of special institutions such as the Joint Coal Board and Coal

Industry Tribunals as "carve outs" that enabled collective bargaining to flourish in tandem with centralised arbitration, as well as favourable industrial policy, trade agreements, training regimes, industrial policies, tariff policies, social security laws and economic development initiatives. These structural legacies combine to leave Australian mining workers with the relatively highest wages by sector in the Australian economy, and with an incapacity to share those material advantages (through formal institutional means) with other worker groups or to use their status to lever their employers out of fossil fuel extraction altogether.

There are powerful limits in the extent to which it is possible to meaningfully bargain over environmental issues within the framework of the *Fair Work Act 2009*, as Schofield-Georgeson and team discuss in Chapter 8 of this collection.[49] The Act includes no environmental objectives in its overall purpose, nor are parties permitted to include enforceable environmental issues in enterprise agreements or include them as an objective of industrial action. Industrial action to shut down electricity utilities, for instance, of the kind proposed by Huber as a strategy for building power to compel low-carbon transformations, is practically foreclosed by the Fair Work Act. Nor are environmental values relevant to work value claims or redundancy processes.

Just cessation: A proposal for reconfiguring just transition to better address the climate crisis

Just transition, as it is currently configured, invites people to expect that it is possible to reorganise society along environmentally sustainable lines through processes that are spontaneous, consensual, and do not involve changing the underlying laws and institutions. In reality, the inheritance of past laws and institutions powerfully constrains the ability of current actors to take these steps in the context of fossil fuel extractive industries. How, then, might just transition be practically reworked so as to restore its conceptual coherence, and enable it to contribute meaningfully as a framework for advancing the cessation of profitable fossil fuel extraction projects, as well as those where the economic logic already supports cessation? As Stein and Carr observe, there is no time to perfect the definition of just transition, and the proposal that follows is not an attempt to fundamentally reconfigure just transition principles to align environmental and labour norms.[50] It is rather an addition to existing guidelines that reflects the imperative to phase out fossil fuel extraction and infrastructure, consistent with the findings of the International Energy Agency. Such a statement could potentially mirror the language used by the UN Secretary-General, who recently called for "no new coal and the phasing out of coal by 2030 in OECD countries and 2040 in all other countries," "ceasing all licensing or funding of new oil and gas," "stopping any expansions of existing oil and gas reserves" and "establishing a global phase down of existing oil and gas production compatible with the 2050 global net zero target."[51]

The addition of a statement along the above lines (a "just cessation" statement) would have the effect of creating firmer boundaries to the ambit of just transition to exclude matters that involve ongoing fossil fuel extraction beyond the prescribed timeframes, or that involve the delaying or undermining of fossil fuel cessation policies. It would also recognise the profoundly asymmetrical responsibilities of different employment relations actors in determining whether fossil fuel extraction continues or is phased out, disrupting the implication that each employment relations actor enjoys an equal (and spontaneous) capacity to assume climate responsibility. The explicit recognition of the centrality of fossil fuel extraction phase out as an objective of just transition would enable incumbent fossil fuel employers and unions to anticipate and accept the inevitability of change. It would give them time to revise their own organisational objectives, investments, relationships and economic and legal dependence on ongoing fossil fuel mining and infrastructure.[52] It would also assist workers, unions and communities to plan and arrange support for transition into different sectors.[53]

Recognising the imperative to phase out one particular industry in light of a non-economic social objective is consistent with the philosophical foundations of labour law.[54] The idea is not wholly novel in Australian history. Following the Second World War, vast numbers of Australian military veterans were moved into employment in a short space of time through Reconstruction programmes in Australia, with 600,000 placed in employment by mid-1946. Although the government provided a special transition allowance to demobilised soldiers to supplement their unemployment benefit, the numbers of people who claimed it were negligible.[55] A range of non-economic forms of support such as training, land settlements, housing loans, vocational training, job placement legal advice and medical benefits were provided.[56] Applicants for education and training were assessed for their capacity to undertake their chosen course and whether there was vocational demand for graduates with the sought after qualification, with no expectation of geographic or vocational continuity on the part of the applicant. Accepted servicemen received free tuition, a living allowance of up to three years, travel expenses and supplementary allowances for books and equipment.[57] The terms of policies were the subject of social dialogue with unions and employers. However, significantly (for a climate change context) tripartite negotiations related to the terms of workers' *new* employment, rather than over *whether* a transition to civilian life itself was necessary, which was taken as given.

Conclusion

The climate crisis requires a holding together of two strands of historical imagination. The first is a continuity story, in which presently existing skills of repair and care are brought to bear in reconfiguring the material world to fit within ecological limits, and accorded new and increased recognition and security for their social importance.[58] The second is a story of radical discontinuity, in which the existing infrastructure of fossil fuel extraction is phased out within a publicly announced and enforced timeframe, where no further licences or permits for fossil fuel projects (or extensions) are granted, profitable fossil fuel resources are left in the ground, and existing workforces are supported to transition into retirement or other industries with reference to their relationship to carbon extraction (rather than the economic decisions of their employers). It is crucial to recognise that we meet this challenge with institutional legal forms that were not built for radical discontinuity. Just transition, as currently conceived, makes most salient the continuity dimensions of labour-climate changes, and it does so primarily because it was created as an expression of the norm of "sustainable development," at a time before it was widely recognised that the cessation of fossil fuel extraction was a *sine qua non* of every other development goal.[59]

As it is presently framed, just transition is vulnerable to being deployed to delay and deny the need for fossil fuel cessation. It also provides cues to parties to look on the reconfiguration of existing labour laws that support fossil fuel extractive occupations as extraneous to the "real" subject matter of just transition. This chapter argues that integrating a just cessation statement to just transition would constitute a helpful first step in tightening the conceptual coherence of the idea for the climate change age. It would also begin to align just transition with what is now known about the ecological preconditions for social stability and flourishing, and thus contribute to aligning just transition with what is, for the ILO and many key theorists of labour law, the essential purpose of labour law.

Reconstituting the basis of labour law around what is known about climate science is not the same as using scientific facts to "shame" workers (a tendency of some strands of environmental discourse that Huber rightly warns against). Rather, it is a modest step towards a material re-alignment between the conditions of different kinds of occupations within labour market structures and the ecological carrying capacity of the planet. Such realignments must be the concomitants to the redistributive policies envisaged by Green New Deal movements and their equivalents across the world. Human solidarity will always overflow and transcend the categories provided for it by law. However, there are powerful ways in which existing legal frameworks position workers against the ecological foundations for collective well-being, including by denying scope for environmental norms in bargaining and industrial action, and by offering relatively more secure material, representational and status conditions for fossil fuel extractive workers. Realigning labour law to better support workers engaged in work that does not deepen the climate crisis, that is "repair and care" work of all kinds, will be a crucial long-term dimension to the achievement of the kind of majoritarian, material climate politics that Huber argues for. To get there, the "deck" upon which work occurs must, like Neurath's boat, be rebuilt at sea, a task that will involve many hands, as well as novel groupings of solidarity and conflict.

Notes

1 Levien, "White Energy Workers of the North, Unite?"
2 Huber, "The Professional Class Vanguard of Climate Justice."
3 Huber, "The Professional Class Vanguard of Climate Justice."
4 ILO, "Just Transition towards Environmentally Sustainable Economies and Societies for All," 4.
5 The UN conception of just transition includes the recognition of the integral place of "gender equality, racial equality and social cohesion; protection of the rights of Indigenous Peoples; disability inclusion; intergenerational equity and young people; the promotion of women and girls; marginalised persons' leadership and involvement in decision-making; and recognition of the value of their knowledge and leadership; and support for the collective climate action of diverse social groups." UN Climate Change Conference UK (COP26), *Supporting the Conditions for a Just Transition Internationally*.
6 York and Bell, "Energy Transitions or Additions?" 40–43.
7 ILO, *Guidelines for a Just Transition*, 4.
8 Ciplet and Harrison, "Transition Tensions," 440.
9 Stevis, *Just Transitions*, 2–3.
10 Mazzocchi, *An Answer to the Jobs-Environment Conflict?*
11 Stevis, *Just Transitions*, 15.
12 The percentage of "unextractable" fossil fuel reserves is particularly high in Australia (where 95 percent of coal, for example, must remain *in situ*). Welsby et al., *Unextractable Fossil Fuels in a 1.5C World*, 233.
13 Leaton, *Unburnable Carbon*; Paterson, "The End of the Fossil Fuel Age?"
14 IPCC, *Synthesis Report*, 53.
15 International Energy Agency, *Net Zero by 2050*.
16 Guterres, "Secretary-General's Video Message."
17 Steffen et al., "Trajectories of the Earth System in the Anthropocene."
18 For instance, the US Willow project will require oil workers until 2053; Australian coal licenses have been granted to mines that will not expire until the end of the twenty-first century. Killoran, "Dirty Dozen."
19 Parry, Black and Vernon, "Still Not Getting Energy Prices Right."
20 Barton, *Fossil Fuel Mineral Wealth and Climate Change Law*, 469, 470.
21 Woodside, *Just Transition*.
22 Kalt, "Jobs vs. Climate Justice?" 1143.
23 Maher, "Energy, Mining and Climate, Fact and Fiction."
24 Tony Maher in Peter Sheldon, *The Ruhr or Appalachia?*, ii.
25 See Carr and Larkin, "Energy Transitions in a Metallurgical Coal Region," Chapter 7 of this volume.
26 Bhaskar, Assadi, and Somehsaraei, "Decarbonization of the Iron and Steel Industry," 758.
27 Campbell, "You Must Be Coking!"
28 Macrogenis, *Just Transition is Possible!*
29 Saget, Luu, and Karimova, "A Just Transition," 480–81.
30 Arora and Schroeder, "How to Avoid Unjust Energy Transitions," 9.
31 Caldecott, Sartor, and Spencer, *Lessons from Previous "Coal Transitions*," 5.
32 Saget, Luu, and Karimova, "A Just Transition," 480–81.

33 Caldecott, Sartor, and Spencer, *Lessons from Previous "Coal Transitions,"* 10.
34 Mercier, *Four Case Studies on Just Transition*, 39.
35 Arora and Schroeder, "How to Avoid Unjust Energy Transitions," 9.
36 Chandrashekeran, *Re-stating Power*.
37 Babic and Dixon, *Decarbonising States as Owners*.
38 Mining and Energy Union, *Rules of the Mining and Energy Union*.
39 Burgmann and Milner, "Ecotopians in Hardhats."
40 Stevis, *Just Transitions*, 8.
41 Carr, "Repair and Care"; Briggs, *Renewable Energy Jobs*.
42 Flanagan, "Just Cessation."
43 Flanagan and Goods, "Climate Change and Industrial Relations."
44 Ghosh, "Rights and Coercion."
45 Historically, the system of "connective bargaining" that prevailed in Australia until 1971 positioned the Metal Trades as the "lead sector" that was the source of flow-on improvements in wages and conditions to the rest of the workforce. Heino, "The Engine of Antipodean Fordism."
46 Mitchell, *Carbon Democracy*.
47 Macintyre, *Australia's Boldest Experiment*, 107–8.
48 Baer, "The Nexus of the Coal Industry and the State in Australia," 196.
49 Schofield-Georgeson, Rawling, and Heino, "Industrial Democracy and Climate Struggle," see Chapter 8 of this volume.
50 Stein and Carr, "Introduction: Design, Material Knowledge and Skillful Practice," in this volume.
51 Guterres, "Secretary-General's Video Message." For an analysis of how equity considerations might be applied to fossil fuel phase-out, see Muttitt and Kartha, "Equity, Climate Justice and Fossil Fuel Extraction."
52 Caldecott, Sartor, and Spencer, *Lessons from Previous "Coal Transitions,"* 14.
53 Stein, "Unresolved Tensions in Green Transitions."
54 Flanagan, "Just Cessation," 393.
55 Macintyre, *Australia's Boldest Experiment*, 325.
56 Macintyre, *Australia's Boldest Experiment*, 318.
57 Macintyre, *Australia's Boldest Experiment*, 329.
58 Carr, "Repair and Care."
59 Daley and Lawrie, *Fuelling Failure*.

Bibliography

Armeni, Chiara and Maria Lee. "Participation in a Time of Climate Crisis." *Journal of Law and Society* 48, no. 4 (2021): 549–72.

Anmol, Arora, and Heike Schroeder. "How to Avoid Unjust Energy Transitions: Insights From the Ruhr Region" *Energy, Sustainability and Society* 12, no. 19 (2022): 1–13.

Australia Institute, *Australian Fossil Fuel Subsidies Surge to $11.6 Billion in 2021–22*. Media release. Canberra: Australia Institute, 28 March 2022. https://australiainstitute.org.au/post/australian-fossil-fuel-subsidies-surge-to-11-6-billion-in-2021-22/

Babic, Milan and Adam D. Dixon. "Decarbonising States as Owners." *New Political Economy* 28, no. 4 (2022): 608–27.

Baer, Hans A. "The Nexus of the Coal Industry and the State in Australia: Historical Dimensions and Contemporary Challenges." *Energy Policy* 99 (2016): 194–202.

Barton, Barry. "Fossil Fuel Mineral Wealth and Climate Change Law: Expectations of Coal Mine Development in a Time of Decarbonisation." *Journal of Energy & Natural Resources Law* 39 (2021): 469–88.

Bhaskar, Abhinav, Mohsen Assadi and Homam Nikpey Somehsaraei. "Decarbonization of the Iron and Steel Industry with Direct Reduction of Iron Ore with Green Hydrogen." *Energies* 13, no. 3 (2020): 758.

Briggs, Chris. *Renewable Energy Jobs in Australia: Stage One.* Report. Sydney: UTS Institute for Sustainable Futures, June 2020) https://assets.cleanenergycouncil.org.au/documents/resources/reports/Clean-Energy-at-Work/renewable-energy-jobs-in-australia.pdf [accessed 5 July 2022]

Burgmann, Verity and Andrew Milner. "Ecotopians in Hardhats: The Australian Green Bans Movement." *Utopian Studies* 22, no. 1 (2011): 125–42.

Caldecott, Ben, Oliver Sartor and Thomas Spencer. *Lessons from Previous 'Coal Transitions.'* Report. IDDRI and Climate Strategies, 2017. https://www.iddri.org/sites/default/files/import/publications/coal_synthesisreport_v04.pdf [accessed 21 November 2022]

Campbell, Rod. "You Must Be Coking! Are New Coalmines OK If They Help Make Steel?" *Crikey* (21 July 2023). https://www.crikey.com.au/2023/07/21/new-coalmines-steel-coke-metallurgical/ [accessed 4 December 2023]

Carr, Chantel. "Repair and Care: Locating the Work of Climate Crisis." *Dialogues in Human Geography* 13, no. 2 (2022): 221–39.

Chandrashekeran, Sangeetha. "Re-stating Power: How States Make, Tame, and Shape Markets." *Geographical Research* 60, no. 1 (2022): 138–51.

Ciplet, David and Jill Lindsey Harrison. "Transition Tensions: Mapping Conflicts in Movements for a Just and Sustainable Transition." *Environmental Politics* 29, no. 3 (2020): 435–56.

Daley, Freddie and Charlie Lawrie. *Fuelling Failure: How Coal, Oil and Gas Sabotage All Seventeen Sustainable Development Goals.* Report. University of Sussex and Fossil Fuel Non-Proliferation Treaty, 2022. https://static1.squarespace.com/static/5dd3cc5b7fd99372fbb04561/t/629621606337cb2779a632f9/1654006125016/FFN_MVSA003+Report+-+Fossil+Fuels+vs.+the+Sustainable+Development+Goals_V4-FA-Screen-Single.pdf [accessed 5 April 2023]

Doorey, David J. and Ann Eisenberg. "The Contested Boundaries of Just Transitions." In *Labour Law and Ecology*, edited by C. Chacartegui. Rochester, NY: Global Law Collection, ARANZADI/CIVITAS, 2022).

Flanagan, Frances. "Just Cessation: How Might the Climate Imperative to Phase Out Fossil Fuel Extraction Reshape the Conception of Just Transition?" *International Journal of Comparative Industrial Relations and Labour Law* 39, no. 3/4 (2023): 393–418.

Flanagan, Frances and Caleb Goods. "Climate Change and Industrial Relations: Reflections on an Emerging Field." *Journal of Industrial Relations* 64, no. 4 (2022): 479–98.

Ghosh, Devleena. "Rights and Coercion: Adivasi Rights and Coal Mining in Central India." In *Dipesh Chakrabarty and the Global South*, edited by Saurabh Dube, Sanjay Seth and Ajay Skaria, 93–104. London: Routledge, 2020.

Guterres, Antonio. "Secretary-General's Video Message for Press Conference to Launch the Synthesis Report of the Intergovernmental Panel on Climate Change." Video message. 2023. https://www.un.org/sg/en/content/sg/statement/2023-03-20/secretary-generals-video-message-for-press-conference-launch-the-synthesis-report-of-the-intergovernmental-panel-climate-change [accessed 3 April 2023]

Heino, Brett. "The Engine of Antipodean Fordism: Australia's Metal Trades Award, 1947–63." *Law and History* 6 (2019): 143–74.

Huber, Matthew. "The Professional Class Vanguard of Climate Justice: A Response to Michael Levien's Review of Climate Change as Class War." *Historical Materialism*, 20 April 2023. https://www.historicalmaterialism.org/blog/professional-class-vanguard-climate-justice-response-to-michael-leviens-review-climate-change [accessed 3 May 2023]

Intergovernmental Panel on Climate Change (IPCC). *Synthesis Report of the IPCC Sixth Assessment Report: Longer Report*, 2023. https://report.ipcc.ch/ar6syr/pdf/IPCC_AR6_SYR_LongerReport.pdf [accessed 5 April 2023]

International Energy Agency. *Net Zero by 2050 – A Roadmap for the Global Energy Sector, 224*. 2021, https://www.iea.org/reports/net-zero-by-2050

International Labour Organization (ILO). *Guidelines for a Just Transition towards Environmentally Sustainable Economies and Societies for All*. 2015, https://www.ilo.org/wcmsp5/groups/public/@ed_emp/@emp_ent/documents/publication/wcms_432859.pdf [accessed 7 December 2022]

Kalt, Tobias. "Jobs vs. Climate Justice? Contentious Narratives of Labor and Climate Movements in the Coal Transition in Germany." *Environmental Politics* 30, no. 7 (2021): 1135–54.

Killoran, Matthew. "Dirty Dozen: Qld Coalmines to Stay Open beyond Net-Zero Deadline." *Courier Mail*, 28 November 2022.

Leaton, James. "Unburnable Carbon: Are the World's Financial Markets Carrying a Carbon Bubble?" Report. *Carbon Tracker*, 2014. https://carbontracker.org/wp-content/uploads/2014/09/Unburnable-Carbon-Full-rev2-1.pdf [accessed 10 April 2023]

Levien, Michael. "White Energy Workers of the North, Unite? A Review of Huber's *Climate Change as Class War*." *Historical Materialism*, n.d. www.historicalmaterialism.org/book-review/white-energy-workers-north-unite-review-hubers-climate-change-class-war [accessed 24 April 2023]

Macintyre, Stuart. *Australia's Boldest Experiment: War and Reconstruction in the 1940s*. Sydney: NewSouth Books, 2015.

Macrogenis, Stavros. "Just Transition is Possible! The Case of Ruhr (Germany)." *Just Transition*, 11 April 2018. https://www.just-transition.info/just-transition-is-possible-the-case-of-ruhr-germany/

Maher, Tony. "Energy, Mining and Climate, Fact and Fiction." *Speech to the Sydney Institute*, 27 November 2019. Mining and Energy Union (MEU). https://meu.org.au/policy-research/energy-mining-and-climate-fact-and-fiction/

Mazzocchi, Tony. "An Answer to the Jobs-Environment Conflict?" *Green Left Review*. 8 September 1993. https://www.greenleft.org.au/content/answer-jobs-environment-conflict

Mercier, Sinead. *Four Case Studies on Just Transition: Lessons for Ireland*. Research Series, Paper No. 15. National Economic and Social Council. May 2020. http://files.nesc.ie/nesc_research_series/Research_Series_Paper_15_TTCaseStudies.pdf [accessed 19 December 2022]

Mining and Energy Union (MEU). Rules of the Mining and Energy Union. Registered 1 December 2023. https://meu.org.au/wp-content/uploads/2023/12/Mining-and-Energy-Union-Rules.pdf

Mitchell, Timothy. *Carbon Democracy: Political Power in the Age of Oil*. London: Verso, 2011.

Muttitt, Greg and Sivan Kartha. "Equity, Climate Justice and Fossil Fuel Extraction: Principles for a Managed Phase Out." *Climate Policy* 20, no. 8 (2020): 1024–42.

Parry, Ian W.H., Simon Black and Nate Vernon. "Still Not Getting Energy Prices Right: A Global and Country Update of Fossil Fuel Subsidies." *IMF Working Paper*, 24 September 2021. https://www.imf.org/en/Publications/WP/Issues/2021/09/23/Still-Not-Getting-Energy-Prices-Right-A-Global-and-Country-Update-of-Fossil-Fuel-Subsidies-466004 [accessed 6 April 2023].

Paterson, Matthew. "The End of the Fossil Fuel Age? Discourse Politics and Climate Change Political Economy." *New Political Economy* 26, no. 6 (2021): 923–36.

Saget, Catherine, Trang Luu, and Tahmina Karimova. "A Just Transition Towards Environmental Sustainability for All." In *The Palgrave Handbook of Environmental Labour Studies*, edited by Nora Räthzel, Dimitris Stevis and David Uzzell, 467–91. Cham, Switzerland: Springer, 2021.

Sheldon, Peter, Raja Junankar, and Anthony De Rosa Pontello. *The Ruhr or Appalachia? Deciding the Future of Australia's Coal Power Workers and Communities*. Industrial Relations Research Centre Report for CFMMEU Mining and Energy. Sydney: UNSW Business School, 2018.

Stanford, Jim. *Employment Aspects of the Transition from Fossil Fuels in Australia*. Report. Canberra: Centre for Future Work, 2021.

Steffen, Will, et al. "Trajectories of the Earth System in the Anthropocene." *Proceedings of the National Academy of Sciences* 115, no. 33 (2018): 8252–59.

Stein, Jesse Adams. "Unresolved Tensions in Green Transitions: Retraining and the Question of 'How'?" *Dialogues in Human Geography* 13, no. 2 (2022): 244–48.

Stevis, Dimitris. *Just Transitions: Promise and Contestation*. Cambridge: Cambridge University Press, 2023.

UN Climate Change Conference UK (COP26). *Supporting the Conditions for a Just Transition Internationally*, 2021. https://webarchive.nationalarchives.gov.uk/ukgwa/20230313132211/https://ukcop26.org/supporting-the-conditions-for-a-just-transition-internationally/ [accessed 26 November 2024]

Welsby, Dan et al. "Unextractable Fossil Fuels in a 1.5C World." *Nature* 597 (2021): 230–34.

Woodside. *Just Transition*, July 2021. https://www.woodside.com/docs/default-source/sustainability-documents/climate-change/just-transition-(july-2021).pdf

York, Richard and Shannon Elizabeth Bell. "Energy Transitions or Additions?: Why a Transition from Fossil Fuels Requires More than the Growth of Renewable Energy." *Energy Research and Social Science* 51 (2019): 40–43.

7

PLACE-SENSITIVE APPROACHES TO COAL TRANSITIONS IN AUSTRALIA

Chantel Carr and Natasha Larkin

Introduction

As one of the world's largest coal exporters, Australia's substantial coal and gas reserves have profoundly shaped national debates around decarbonisation and climate policy. The deep integration of extractive sectors within Australia's regional communities has created a complex landscape for change that demands careful consideration of diverse perspectives and experiences.[1] While subnational state governments have made progress on transitioning domestic electricity generation away from coal, decades of national policy instability reflect broader political contestation.[2] Within the context of mounting domestic and international pressure, the election of the Albanese Labor government in 2022 marked an accelerated national commitment to decarbonising difficult-to-abate industrial sectors. Many of these sectors are also concentrated in regional Australia, and this political shift has re-framed the narrative from "jobs versus environment" to one of regional renewal, delivered through industrial-scale renewable energy and decarbonisation initiatives. Yet researchers have long pointed to the uneven outcomes of centralised policymaking for regional workers and communities.[3]

Over 80 percent of Australian coal production is exported, predominately to Asian markets.[4] Despite federal and subnational state policy oversight, the future of export coal remains opaque and highly politicised, with both politicians and policymakers reluctant to address – or even discuss – phase-out pathways. An absence of such policy, whether intended or otherwise, effectively leaves decisions about the future of Australian coal exports in the hands of global capital and trading partners. The implication is that shifts in the global political economy of coal will have profound economic, social and political

DOI: 10.4324/9781003512240-10

consequences in Australia, particularly in export-oriented communities, yet this is rarely the subject of public or political discussion.

The legislation of a national Net Zero Economy Authority in 2024 brought some measure of protection for workers in regions hosting coal-fired power generation[5]; however justice outcomes for export-exposed workers and communities remain uncoordinated at a national scale. The language and framework of a "just transition"[6] has been embraced and pursued by some powerful actors, including segments of organised labour, scholars and policymakers. However research has identified that at the local level, some coalmining communities view the concept of a just transition as a direct threat to their identity, livelihood and deep connections to place.[7]

Meanwhile, influential voices in politics, media and industry have promoted narratives that downplay Australia's role in global climate change. These narratives have gained varying degrees of acceptance within communities where researchers have found more nuanced perspectives at the community level. In coal regions, views about the future of the industry can often defy polarised national dialogues that favour either immediate industry closure or indefinite continuation.[8] Building on research that emphasises the importance of place-based analyses of energy transitions,[9] in this chapter we examine the experiences and perspectives of metallurgical coal worker households in the Illawarra region of New South Wales (NSW). Our analysis contributes to calls for a more nuanced understanding of worker and community responses to transition.[10]

Human geographers have long advocated for the household as an important site for research,[11] and this is certainly the case for examining the social dimensions of industrial transformation. Households are key spaces where decisions about working futures are made during periods of economic change, and where the outcomes of industrial restructuring are acutely felt.[12] A focus on households offers a different entry point for examining the complex relationship between carbon-intensive regions and climate change, recognising that while workers have some agency in career choice, their decisions are also shaped by economic opportunities, household obligations, and the structural limitations of regional labour markets. By attuning to the intersection of different industries within a household, the allocation of unpaid labour, and community relations beyond individual (predominately male) workers, we can better understand the constraints and considerations that shape livelihoods in carbon-intensive sectors and regions. This invites a more careful conversation about how decarbonisation will reconfigure working lives, impacting on households and communities in the process.

In this chapter we begin with an overview of the political economy of Australian coal, drawing attention to how regional trajectories are shaped by many factors: material distinctions in coal reserves, global markets, and national and subnational state policies. We then outline the shifting social and economic dynamics in the Illawarra region of NSW, providing context for our empirical data, and an analysis of the perspectives of coal worker households in the region. Here, we draw out three themes that emerge from the interviews: the daily realities of working within export-oriented underground coal mining; how ongoing economic precarity is experienced and negotiated by workers and households; and different visions for the future of the industry. Ultimately, these grounded insights highlight some critical but often overlooked dynamics within Australia's energy transition. They point to the need for place-based sensitivity and clearer transition policy pathways for export-oriented commodities, to better support workers, households and communities across diverse regions.

Locating Australia within a global political economy of coal

Given coal has been so highly politicised in Australia, a great degree of care is required in how it is discussed. Two key distinctions structure Australia's coal sector: the difference between thermal and metallurgical coal; and the difference between domestic and export coal markets. These distinctions shape both how Australia is positioned within a global political economy of coal, and the future trajectories of coal-producing regions, as determined through their different relationships with global markets.

Metallurgical coal is used for steelmaking, which contributes between eight and ten percent of the world's carbon emissions.[13] Australia is the world's largest metallurgical coal exporter, with reserves concentrated in the geological basins of central Queensland, and the Hunter and Illawarra regions of NSW.[14] These regions are deeply integrated with global metallurgical coal markets, with nearly all Australian metallurgical coal exported to major trading partners including India, Japan and South Korea.[15] The implication is that communities in metallurgical coal regions are particularly vulnerable to global market forces, and decisions made in boardrooms are at an increasing distance from those who depend on the sector. While some (predominantly European) steelmakers are pursuing alternatives to replace metallurgical coal in the steelmaking process through methodologies such as Direct Reduced Iron (DRI) and electric arc furnaces, at the time of writing commercial viability at scale remains a work in progress.[16] This places Australia's metallurgical coal production (and the regions that mine it) on a trajectory heavily influenced by global markets, and by the commitments international trading partners make to decarbonising steelmaking.[17]

Meanwhile thermal coal – used to generate energy in coal-fired power stations – presents a different set of dynamics. Much of the national public discussion on coal in the last two decades has focused on domestic demand for thermal coal. This focus makes sense, as coal-fired power generation has been in decline in Australia due to a number of factors. One key enabler of this relatively rapid transition is national and sub-national policy settings that have supported the decarbonisation of the electricity grid, impacting on domestic demand for thermal coal. An unprecedented rise in domestic solar in Australia has meant that by November 2024, four million – or one in three – homes were fitted with rooftop systems. Over the last decade, investment in larger-scale renewable projects including utility-scale solar, onshore wind and hydro-electricity projects has also accelerated.[18] Much of this has been driven by subnational state policies, where climate politics have arguably been less contentious than the national scale. The withdrawal of global capital from coal-fired energy generation has also been a contributing factor, where the risk of stranded assets and the high maintenance costs associated with ageing infrastructure has further driven transition policy.

While much of the public focus on coal transitions has been focused on domestic thermal coal consumption, in 2024 75 to 85 percent of Australia's thermal coal was exported for energy generation in other countries, including Japan, China, Taiwan and South Korea.[19] The implication is that – as for metallurgical coal – regions producing thermal coal for global markets are exposed to both the policy decisions of international trading partners and fluctuations in global capital investment. The degree of exposure extending beyond national borders is shown in Figure 7.1, which shows a dramatic rise in metallurgical and thermal coal production over the past 40 years. While domestic consumption of coal has remained steady and then dropped off over the last decade with the introduction of renewables, coal exports have continued to expand until very recently, now accounting for 80 to 90 percent of Australia's total coal production. This illustrates very clearly a misalignment between predominant national discussions of coal phase-out that focus on domestic energy generation, and the lived experiences of workers in export-oriented coal regions.

Among the important factors for Australia's growth in coal exports over recent decades is the method for calculating emissions targets under the 2015 Paris Agreement.[20] Under these arrangements, Australia is responsible for the Scope 1 and 2 emissions associated with the production of coal, but not the substantially greater Scope 3 emissions produced in the burning of coal, which fall to the country where it is used. This framework has enabled Australia to focus its efforts on the relatively "low-hanging fruit" of reducing domestic emissions, while actively expanding coal exports. Growing demand for Australian coal has been shaped by several global factors, including rapid economic growth across Asia that has increased regional demand for both thermal and metallurgical coal. Meanwhile, geopolitical disruptions – including most recently, Russia's invasion of Ukraine – have led to increased demand for thermal coal in other global regions.[21] These conditions have fuelled an extended boom period for Australian coal exports.

In terms of coordinated policymaking, resource extraction is governed by a complex combination of federal and State laws and policies under Australia's

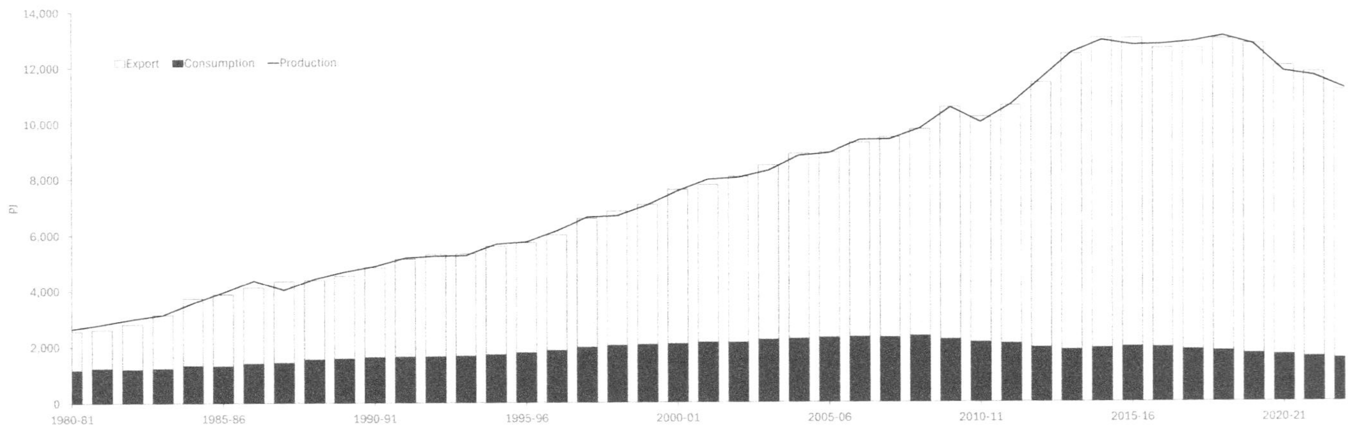

FIGURE 7.1 Australian coal production, domestic consumption and exports, over time. Source: Authors, adapted from DCCEEW Australian Energy Statistics 2024, Tables C1 and J1.

federated constitutional arrangements. Coal reserves are legally the property of subnational state governments and coal mining is regulated by state laws, with some federal and state oversight under environmental protection legislation.[22] Here, the materiality of the coal types interacts with state policies and agendas. For example, the state of Victoria had an ambitious agenda to transition from brown coal used only in domestic coal-fired power generation to meet its net zero targets, whereas the NSW and Queensland governments continue to generate considerable royalty revenue from their thermal and metallurgical black coal exports. Queensland's progressive coal royalties system introduced in 2022 has earned the state AUD$9.4 billion in revenue since its inception, and in 2022–2023 NSW earned AUD $4.5 billion in coal royalties.[23] These substantial revenue streams, combined with the absence of responsibility for Scope 3 emissions, create a powerful disincentive for governments to intervene in coal export markets.

The combination of complex political dynamics and a lack of a clear national policy direction creates significant uncertainty for communities in export-oriented coal regions. Coal workers and their households are left to piece together possible futures from a fragmented landscape of corporate press releases and internal briefings, media speculation, political announcements, and informal localised knowledge-sharing. That is, the absence of a coordinated policy direction leaves communities to interpret mixed signals about their future prospects. While export-oriented coal workers have in recent years experienced sustained investment, growing production and industry rhetoric about long-term market demand, this reality contrasts sharply with the national dialogue around transitions away from coal-fired power generation. More critically, it also overlooks the fact that their livelihoods increasingly depend on the opaque risk calculations of global coal capital, which can divest from assets with little notice or regard for community impacts.

The outlook for coal exports faces additional uncertainty as international trading partners move forward with their own committed emission reduction targets, and Australia comes under increasing international and domestic pressure to reduce its economic dependence on fossil fuel exports. Figure 7.1 shows a recent peak in Australian coal exports, and government economic modelling continues to suggest exports will decline from this point.[24] At the time of writing, no firm national or state policy settings are seeking to curtail coal exports, and some key actors are actively seeking alternative markets – particularly India, which has a net zero by 2070 target.[25] This disconnect between Australia's continued pursuit of coal export markets and its climate change commitments reveals a policy approach that overlooks meaningful decarbonisation beyond domestic emissions. It also leaves export-oriented workers and communities exposed to global capital flight, with very little capacity for sufficient forward-planning of their futures.

Read together, the geographies of different coal types embed specific regions in distinct markets – metallurgical and thermal, domestic and export. As Australia pursues its net zero commitments, these resource geographies will continue to shape uneven regional transition trajectories. The phase-out of low-quality brown coal has focused early transition attention on Victoria's Gippsland region, while central Queensland and the Hunter regions of NSW mine higher quality black thermal *and* metallurgical coal for both domestic use and export. This places these regions on trajectories heavily dependent on global markets and the decarbonisation policy settings of international trade partners. The Illawarra region faces its own distinct vulnerabilities: while three million tonnes of high-quality metallurgical coal supplies the Port Kembla steelworks annually, the rest is exported – yet the region has received little attention in coal transition policymaking. As coal capital shifts offshore and decision-making becomes increasingly distant from those dependent on it, such regions become even more vulnerable to complex global market dynamics. In the next section, we examine how this global political economy of coal interacts with local social and economic dynamics, before turning to our case study of metallurgical coal households in the Illawarra region of NSW.

The Illawarra region: Historical context and contemporary transformations

To situate our empirical data, we begin with a brief overview of the Illawarra region. Declared a region by centralised policymakers in the 1960s,[26] what we now call the Illawarra is the traditional custodial lands of First Nations groups, including Dharawal, Wodi Wodi and Yuin people, who have cared for the land and sea for thousands of years.[27] The region's distinctive geography is shaped by the Pacific Ocean to the east, and a tall escarpment to the west, where three substantial seams of black coal have attracted mining activity since early colonisation. Located just 80 kilometres south of Australia's largest city and economy, Sydney, the region today has a population of around 313,000, with most people living within five kilometres of the ocean.[28] The region's largest city, Wollongong, sits at the centre, with the industrial hub of Port Kembla immediately to its south. Here a vibrant multicultural community lives alongside Australia's largest steel manufacturer, BlueScope, and the industrial-infrastructure complex of Port Kembla, through which the region's high-quality metallurgical coal is exported to Asian markets.

By the early 1900s the Illawarra's economy was dominated by coalmining, but this was overtaken by steelmaking from the mid-twentieth century onwards.[29] At the height of Australian manufacturing in the 1960s, around 22,000 steelworkers (one in four households)[30] and approximately 20 underground coal mines shaped the region's industrial character. The steelworks drew post-war European migrants who settled around the industrial centre.[31] From the early 1980s, waves of industrial restructuring brought a tumultuous two decades of economic instability and significant unemployment for the region.[32] Youth unemployment was particularly high compared to state averages, prompting state and regional policymakers to pursue economic diversification away from the region's industrial identity. Throughout the 1990s and beyond, a focus on tourism and service sectors emphasised the region's natural assets and sought to bring tourists, university students and commuting professionals into the region.[33] Reflecting these policy efforts, today the region's economy combines heavy industry concentrated around Port Kembla, four operational underground coal mines owned by global capital, and a growing service sector. The largest employers by sector are health, retail, tourism, construction and higher education.[34] Some 26,000 people commute daily between the Illawarra and nearby Sydney.[35]

This shift from industrial heartland to a diversified economy has shifted the region's socio-economic dynamics. Most coastal suburbs now have a concentration of professional commuters in gentrified areas, where house prices exceed the national average. Meanwhile, skilled trades workers and industrial labourers tend to be clustered – though not universally – in suburbs adjacent to key industrial precincts in the region's south. Distinct pockets of socio-economic disadvantage remain across the region. Reflecting national trends,[36] accelerated internal migration from major metropolitan centres (predominantly Sydney) into the region during and immediately after the Covid-19 pandemic has intensified a pre-existing housing shortage, particularly affecting renters and those in need of social housing. These patterns reflect how the Illawarra's geography, resources and economy combine to deeply connect the region to the global and wider domestic economy.

The region has a long history of labour, environmental and social activism. In particular, local branches of the union movement have a history of fighting for better conditions for workers in the coal and steel industries.[37] Community-based activism has also been prominent, including local contests focused on preserving places of significance to First Nations people, and protecting the local environment from industry and over-development. In recent years, community activism has mobilised in opposition to applications to extend existing underground coal mines. Some mining operations extend beneath the Sydney water catchment, while others push up against residential housing and the protected Royal National Park. In our own empirical research, we have found that opposition to mining in the region tends to reflect both local and global climate change concerns, strengthened by an influx of new residents whose economic lives are not tied to the region's heavy industries. The absence of coordinated planning for the future of Australia's export coal industries leaves both coal worker households and the broader community unclear about the future of both mining and steelmaking in the region.

In what follows we turn to our empirical data, exploring how metallurgical coal workers and households navigate daily life and uncertain futures in a changing industrial landscape. Our analysis draws primarily on 32 open-ended interviews conducted between 2021 and 2023 with coal workers and their households, primarily partners and young adults. Interviews lasted between 60 and 150 minutes and were conducted at participants' homes, public libraries and online (the latter during Covid-19 lockdowns). Participants were predominantly permanent employees working across different mines in the Illawarra, though many had experience as contractors in the sector prior to gaining permanent employment. The age range of participants spanned from late twenties to retired workers with over four decades of industry experience. These household interviews were supplemented with ten interviews to gather key stakeholder perspectives from local politicians, industry and regional union representatives, and mine training and rescue personnel. Qualitative data was supported by contextual analysis, including Australian Bureau of Statistics (ABS) data, analysis of federal and state policy documents and publicly available company reports, and ongoing participation in numerous stakeholder forums about the region's industrial future. Throughout this fieldwork we sought to maintain an inclusive methodological approach, engaging with diverse actors and perspectives regarding the region's coal industry. Our analysis focused on three interconnected dimensions of coal worker households' lived experiences: the daily reality of working in export-oriented underground metallurgical coal mines, how this shapes household relations, and how households envisioned their futures amidst broader regional transformations.

Export-oriented underground coal mining: An overview of labour process

Underground coalmining in the Illawarra has a distinct culture shaped by both the physical geography of the region's coal resources and the history of mining in the region. Coalmining was key in establishing many of the region's nineteenth century settlements that became suburbs, closely linking the region's postcolonial industrial, social and spatial history. The proximity of coal mines to residential areas means that coal workers tend to live within the region, enabling them to participate in daily household and social life. This is distinctly different to the "fly-in-fly-out" (FI-FO) culture of mining in regions like central Queensland and Western Australia, where work rosters of rotating weeks regularly separate workers from their families and communities for long periods of time.

Contemporary coalmining is diverse and highly automated, departing significantly from historical images of young men with picks and shovels at the coal face. Underground, there are distinct hierarchies, providing clear reporting lines that are designed to manage risk: production workers report to

Deputies, a statutory role regulated by the state under specific mining Workplace Health and Safety (WHS) legislation.[38] Shift trades (primarily electricians and mechanical fitters) install, inspect and repair production technology "on-the-fly." An Undermanager – also a statutory role – coordinates all mining operations for a shift. Other named statutory roles include Mining, Electrical and Mechanical Engineers, and Engineering Managers. Other underground roles include those responsible for extracting and managing gas, involving highly specialised drilling and ventilation expertise. Above ground, workers may be engaged in maintenance trades, undertaking larger, planned overhauls of underground equipment (though this is increasingly outsourced). There are also roles in administration, engineering, legal and financial services, human resources and other forms of management – underground workers report that such roles have ballooned in recent decades.

In the push to increase production, most mines in the Illawarra use the longwall method of mining: a high-powered cutting system 300 metres wide that works back and forth cutting across the coal face, while a system of automated conveyor belts moves coal to the surface. The presence of large, fast-moving equipment in confined spaces mean the conditions are notoriously dangerous, demanding all underground workers remain highly attuned to a multitude of risks. Beyond the risk of roof falls, the mines in the Illawarra are highly gaseous – predominately producing methane – with the ever-present risk of explosion if the gas is not extracted ahead of cutting. Above-average pay rates reflect the risk associated with underground coalmining. Historically, high levels of unionism and industrial relations disputes have secured improved pay and conditions over time. This relationship between coalmining, remuneration and risk was succinctly described by Participant 18: "It's a lot of money. But the thing about it, when I say it's a lot of money, the dangers go with it."

Historically, work in coal mines in the region was intergenerational, with children following their fathers into the mines.[39] However this is not a distinct characteristic of modern coalmining in the Illawarra – our interviews with households pointed to far fewer intergenerational connections with the coal industry. Many participants indicated that they did not want their children to follow in their footsteps as the work was difficult and precarious. Participant 14 pointed to the impacts of the highly physical work over a lifetime:

> But it's a tough job, it's a dirty job, and it's a backbreaking job, and if you're working on crews underground, it's not an easy job. You don't see a lot of old coal miners underground anymore.

Yet despite the demanding and dangerous conditions, many participants suggested that they had no desire to do other work. Participant 19 explained his enjoyment of a work culture where sharp distinctions could be drawn with the demands of contemporary knowledge-based work: "It's that camaraderie as well, in a crew, having a laugh and a joke, and when you're underground, there's no phones, there's no internet, there's nothing."

Given the conditions described above, it is clear that coalmining has a distinct embodied experience that extends to the organisation of work. Workers in Illawarra-based mines are rostered on ten or 12 hour shifts usually for three to four days at a time, and many have experienced a multitude of shift changes over their careers. Underground workers are required to be punctual for every shift: crews leave at very precise times from the surface, traveling up to 15 kilometres underground to reach the longwall, and do not return until the end of the shift. Personal protective equipment and a self-contained breathing apparatus – up to 15 kilograms in total – is carried by each person working underground. There is no mobile reception and conditions are almost always confined, but also vary depending on where work is carried out: from hot to very cold, dark to well-illuminated, and often very wet, as the coal is hosed down as it is cut and transported, to reduce dust.

Unsurprisingly given these conditions, unions have played a central role in this industry over centuries, but since the industrial reforms of the Howard government era (1996-2007), the sector is increasingly characterised by labour hire, contract workers and reduced unionisation. The fieldwork took place immediately prior to the passing of federal "same job, same pay" legislation,[40] and workers spoke of working in a crew alongside another worker with a different shirt from a labour-hire company, doing the same work, but on different pay and conditions. They spoke of the way this undermined a sense that they were working together as equals, in work that heavily relies on coordination and mutual support for safety. There was also a sense that over-reliance on inexperienced contract workers in times of boom was unsafe. This apprehension extended to workers who were trained in other regions, where safety and risk were sometimes described as less of a concern. While such insights could be read as parochial, they point to a highly contextual and geographically specific labour process, where distinct safety and risk management cultures have evolved over time.

Navigating precarious work: A household perspective

For workers with partners or families, it is at the household level that decisions are made around work – paid and unpaid. The experiences of coal worker households in the Illawarra diverge from more remote mining communities such as those described by Gibson-Graham in central Queensland, where women were left with few economic possibilities. Here, a highly diversified economy means work opportunities are available across a broad range of industries, so juggling two working lives with care responsibilities worked in several ways. In household interviews, some partners (all women) chose to work in lower-paid sectors such as retail or administration, where work was plentiful and they could maintain flexibility and work around their partner's shiftwork in the coal sector. In other households, coal workers spoke about choosing certain shift patterns so they could be there for their families – to see their children in the evening, to be present at weekend sport, and to care for children while their partner worked. This was often enabled by the specific shift patterns and relatively good remuneration of mining, as described by Participant 3:

> The best thing about this career and this industry is that I can work four days, then switch off and I've got three days at home with the family. Yeah, I don't take that for granted.

Participants also spoke of the predictability that rostered work offered them, allowing them to participate more fully in family life and actively volunteer in the community. One consistent theme that emerged from interviews was the equitable sharing of care, with many coal workers reporting sole care of small children on their non-rostered days. As Participant 21 explained: "[My partner] basically works on my days off. We sort of share parenting, I suppose. I have the kids on my days off and obviously on her days off she's got the kids …" This points to the way paid and unpaid work is navigated at the household level, and not at the individual worker level – where industrial relations and broader social dialogues around work are often anchored. Participants also spoke of engagement in a wide range of volunteer and community activities, from participating in social groups such as organised surfing competitions, to coaching children's sporting teams, volunteering on suicide prevention hotlines, and engaging in local environmental campaigns. These dynamics have been to date largely overlooked in transitions discourse, yet they point to how households have been able to make sectoral conditions work in their favour, creating unexpected social benefits that make the work distinctively attractive compared to other employment options.

While such roster arrangements create valuable opportunities for family and community life, they exist within a broader context of corporate power dynamics that do not often favour workers. Coal workers spoke about how employers regularly negotiated changes to shift patterns, impacting not just themselves, but the work of partners and patterns of family life. Beyond changes in shift patterns, households are left to navigate cyclical precarity in a boom/bust industry dominated by global commodity markets. Forced redundancy is a longstanding central characteristic of coalmining work, although this has intensified in recent decades as ownership has shifted from a single vertically integrated operator (BHP) to a mix of global corporate resources entities, some more economically secure than others. This precarity was described by Participant 12 in the following way: "People used to say, you're not a true coal miner until you've been retrenched! … it can change so quick."

These experiences of redundancy and insecurity are felt at the household level, and were explained as stressful experiences. This was described by the partner of a coal worker, Participant 2, in the following way: "At the end of the financial year, they'd always threaten to fire everyone and change everyone's job, and that constant not knowing."

To address household stress in times of precarity, participants spoke of multiple strategies for preparing for potential redundancy – partners taking up work or increasing hours in anticipation, workers upskilling into more secure roles in the mine (from production workers to Deputies or through different certificates in areas like rescue and WHS), and workers holding onto trade qualifications and licencing outside the industry "just in case." Workers with longer careers in the industry spoke about the need to pay off mortgages when times were good, and how they had advised younger or more inexperienced workers not to over-capitalise in "boom" periods, as there would definitely be a "bust" coming around the corner.

Many workers reported trying to leave the industry at various times of their own accord, or being forced to find other work during downturns. It is important to note that they often returned to coalmining afterwards driven by economic necessity – with many finding their qualifications insufficient to maintain comparable living standards in other industries. We found the experiences of many participants echoed research findings on redundancies in other industrial sectors including steelmaking and automotive manufacturing: that workers draw on their informal networks to find alternative work, and that older workers without trade qualifications were most at risk of being unemployed.[41] Given this context of employment precarity and the uncertain future of metallurgical coal in the region, our research explored how households understood their economic futures in the context of decarbonisation. What emerged was a striking diversity of values, approaches and experiences among participants.

Accounting for divergent views on the future of coal

Illawarra coal workers and their households face significant uncertainty about the future of the sector in the region, with no clear policy direction to guide their future planning. As a result, many are left to hedge their bets on the sector's future at the household scale, cobbling together company announcements, reports of global market dynamics, and commentary from the media, politicians and labour institutions, each of which are politicised in their own way. As a result, participants held varied views about the future of the metallurgical coal industry. Where some participants saw the industry in the region had only a few years to continue, others like Participant 7, felt it had decades of viability: "I think with metallurgical coal that might push that out to 35, 40 years, potentially."

While most participants expressed confidence in their own immediate job security, their outlook for future generations was markedly different. Almost all participants with younger children expressed their desire for their children to work in other sectors, given their own lived experiences of labour insecurity. As Participant 1 reflected: "It's not something that I'd be looking for my kids to start a career in at this stage. But I think it'll probably see me out. Possibly I've got twelve years left." This relative confidence about short-term job security stemmed largely from participants' identity as *metallurgical* coal miners. They emphasised how metallurgical coal remained essential to current forms of steelmaking globally. Some participants went further, challenging the notion that their jobs were complicit in climate change by highlighting the central need for steel – and therefore metallurgical coal – in renewable energy infrastructure.

Participants also expressed diverse views on climate change and its relationship with the industry. While some did not believe that coalmining had negative environmental impacts, others expressed personal discomfort about working in resource extraction, but saw it as the best available option for supporting their families. This complexity was particularly evident in the case of Participant 17, who intended to become an environmental activist in retirement: "When I finish my mining career, I'll probably go the other way because, like I said to you, I don't agree with the processes and what they do. It's very bad for the environment." This perspective challenges (often progressive) narratives that portray coal miners as indifferent to climate change, or that seek to frame employment as primarily a moral rather than an economic choice. In a region with historically limited opportunities for those without formal qualifications, coalmining enables some households to maintain a stable standard of living. More recently, the rising cost of living and lack of housing affordability has intensified in Australia creating additional pressures. This is particularly the case in the high-amenity coastal suburbs of the Illawarra, where internal migration has driven costs even higher. In this context, transitioning to lower-paid employment has become increasingly unfeasible for families dependent on mining income.

In making sense of the industry's future, participants described the multitude of strategies and sources they drew from, including company briefings, national and local media, informal discussions with other workers and analysing trends in global commodity markets for metallurgical coal online. In general, participants demonstrated sophisticated understandings of how global market factors intersect with local forces such as state planning regimes, which have recently made mine extensions more challenging due to environmental impacts and localised opposition. The changing social dynamics of the region also shaped how coal miners understood their working futures. As Participant 21 observed: "Everyone started moving in from out-of-town, saying, 'Why the hell is there a coal mine here?' It's like, well, news flash, this whole area was a coal mine!" This comment reflects some growing tensions between the region's industrial history and its emerging identity post-Covid, which is more closely tied to the natural assets of the escarpment and ocean, and the rise of a substantial professional and service economy.

Despite varying views on timelines, many participants shared a view that no single factor will bring about the end of the coal industry in the Illawarra, but rather, it is going to be "death by a thousand cuts," from intersecting forces. Mine owners face increasing difficulty securing licence approvals, particularly for operations located under the water catchment and pushing up against the National Park. The region's demographic shift has introduced a growing professional class who have no economic connection to the region's heavy industries and highly value its natural amenity, leading to opposition to industrial expansion. Meanwhile, global investment in carbon-intensive sectors such as coal is declining, leaving operations increasingly dependent on less financially stable capital, with high debts and less resilience through downturns.

While our research was conducted in an extended boom period characterised by labour shortages where experienced workers had some agency in negotiating conditions, history suggests that this will not be sustained. Recent developments support this view, with the closure of Russell Vale mine due to operational safety breaches, and Australian resources operator South32's sale of two Illawarra assets to a global consortium, GM[3], primarily funded through Singapore. Collectively, these factors suggest the trajectory for the coal sector within the Illawarra is much more likely to involve mine-by-mine closures, rather than a coordinated regional transition.

Conclusion

Our empirical research demonstrates the need for more nuanced conversations about coal transitions, attuned to place-based specifics at the regional scale. While media and policy attention has understandably focused on thermal coal regions facing a capital-led withdrawal from coal-fired power generation, regions like the Illawarra – dominated by export-oriented metallurgical coal – face distinctly different challenges. These regions, sitting outside dominant transition narratives and newly legislated policy initiatives such as the Net Zero Authority, are highly vulnerable to the risk of an uncoordinated and unplanned end to their industry. The impacts are likely to be uneven within regions, with workers holding formal qualifications and transferable skills more readily transitioning to new and alternative industries compared to those whose expertise is primarily industry-specific.

Labour precarity, driven by global export markets, has long been a defining feature of coalmining, with impacts felt most acutely at the household level. Adopting a household lens in transition planning reveals important perspectives on the social dynamics of industrial transformation that are often overlooked in current policy approaches. As Fairbrother and colleagues observed in relation to the end of coal-fired power generation in Gippsland,[42] partners and other household members should be included in support measures such as employment guidance and training packages. Likewise, evidence from historical experiences of industrial transformation should inform contemporary responses – where research has found that structural adjustment programmes have often failed to deliver real outcomes for workers, who ultimately end up relying on local networks to secure employment. While avoiding the individualisation of responsibility that has characterised labour relations in Australia for some time, transition planning must recognise and work with these local and relational dynamics.

Perhaps most importantly, our research reveals how working in the Illawarra's coal sector has facilitated positive forms of household and community life, through shared-care arrangements, community engagement and volunteering. This finding challenges simplistic moral narratives about coal workers and communities.[43] Rather than viewing coal workers – through the lens of individual choice or cultural identity – as part of the problem, we seek to understand their positioning within broader structural constraints. Illawarra coal workers and their households navigate their own personal circumstances in a context shaped by global capital flows and a complex policy landscape. This perspective suggests the need for transition approaches that recognise both the distinctive place-based vulnerabilities and the valuable social arrangements that have evolved around mining work, while also acknowledging the structural forces that constrain workers' agency over their working futures.

Notes

1 Connor, "Energy Futures"; Ey and Sherval, "Exploring the Minescape"; Fleming et al., "Understanding the Resource Curse"; Goodman and Worth, "The Minerals Boom"; Weller and Beer, "State Structures."
2 Baer, "The Nexus of the Coal Industry and the State."
3 Sheldon et al., "The Ruhr or Appalachia?"; Snell, "Just Transition?"; Weller, "The Geographical Political Economy"; Weller, "Just Transition?"
4 IEEFA, 2024, "Australia's Coal Export Market."
5 Colvin, "Contextualizing Coal Communities"; Rainnie and Snell, "The Australian Net Zero Economy Authority."
6 ILO, *Guidelines for a Just Transition*. See also Flanagan, "A Transition with Teeth" in this volume.
7 MacNeil and Beauman, "Understanding Resistance"; Edwards et al., "Towards a Just Transition"; Cha, "A Just Transition for Whom?"; Della Bosca and Gillespie, "The Coal Story."
8 Colvin, "Contextualizing Coal Communities"; Colvin and Przybyszewski, "Local Residents' Policy Preferences."
9 Atkins, "The Geographies of 'Stranded Communities'"; Colvin and Przybyszewski, "Local Residents' Policy Preferences"; Weller, "The Geographical Political Economy"; Snell, "Just Transition?"; Fairbrother, "When Politics Meets Economic Complexity."

10 Colvin, "Contextualizing Coal Communities."; Edwards et al., "Towards a Just Transition."
11 Here we especially look to build on Katherine Gibson's work in Central Queensland's remote mining communities in the 1990s (see Gibson, "Hewers of Cake"; Gibson-Graham, *The End of Capitalism*), along with other important work on households and regional economic change within and beyond geography, including in the Australian context, Head et al., "Zones of Friction"; McDonald et al. "Mining Work, Family and Community"; Bamberry, "Households and Citizenship."
12 Head et al., "Zones of Friction."
13 IEA, *Emissions Measurement.*
14 Geoscience Australia, "Coal."
15 DCCEEW, "Resources and Energy Quarterly, Dec 2024."
16 See Vogl, "Steel Beyond Coal" for a comprehensive overview of progress on the decarbonisation of steelmaking internationally, with a European focus.
17 Auger et al., "The Future of Coal Investment."
18 Parkinson, "Wind and Solar Smash Output Records."
19 DCCEEW, "Resources and Energy Quarterly, December 2024."
20 United Nations, "Paris Agreement."
21 DCCEEW, "Resources and Energy Quarterly, September 2022."
22 DCCEEW, "Coal and Coal Seam Gas – Regulation."
23 Queensland Treasury, "State Budget 2024–25."
24 DCCEEW, "Resources and Energy Quarterly, December 2024."
25 See Hughes et al., *Coal and Gas Futures* for a comprehensive overview of Asia-Pacific demand, including India.
26 Eklund, *The Making and Breaking of Port Kembla.*
27 Donaldson et al., *A History of Aboriginal Illawarra.*
28 ABS, "Illawarra.".
29 Eklund, *The Making and Breaking of Port Kembla.*
30 Eklund, *The Making and Breaking of Port Kembla.*
31 Eklund, *The Making and Breaking of Port Kembla.*
32 Burrows, "Unemployment in the Illawarra."
33 Advantage Wollongong, *Wollongong: A City Transformed.*
34 ABS, 2021.
35 Deloitte, *The Decentralisation of Work in the Illawarra.*
36 Deloitte, *The Decentralisation of Work in the Illawarra.*
37 Eklund, *The Making and Breaking of Port Kembla*; O'Brien and Burrows, "Assessing the Effectiveness."
38 For example, in NSW see Work Health and Safety (Mines and Petroleum Sites Regulation) 2014.
39 See also Dahlgren, *Digging Deeper*; Della Bosca and Gillespie, "The Coal Story.".
40 *Fair Work Legislation (Closing Loopholes) Act 2023.*
41 O'Brien and Burrows, "Assessing the Effectiveness."; Barnes and Weller, "Becoming Precarious?" See also Barnes, "From Occupational Steering," in this volume for a related discussion of occupational citizenship.
42 Fairbrother et al., "Jobs and Skills Transition."
43 Huber, *Climate Change Is Class War.*

Bibliography

Advantage Wollongong. Wollongong: A City Transformed. Submission to the Inquiry into the Australian Government's Role in the Development of Cities, 2017.

Atkins, Ed. "The Geographies of 'Stranded Communities' in Energy Transitions." *Progress in Human Geography* 48, no. 6 (n.d.): 614–35.

Auger, Thomas, Johannes Trüby, Paul Balcombe and Iain Staffell. "The Future of Coal Investment, Trade, and Stranded Assets." *Joule* 5, no. 6 (2021): 1462–84.

Australian Bureau of Statistics (ABS). "Illawarra – SA4." 2021. https://abs.gov.au/census/find-census-data/quickstats/2021/107

Australian Department of Climate Change, Energy, the Environment and Water (DCCEEW). Australian Energy Update, 2024a. https://www.energy.gov.au/publications/australian-energy-update-2024

Australian Department of Climate Change, Energy, the Environment and Water (DCCEEW). Coal and Coal Seam Gas – Regulation, 2021. https://www.dcceew.gov.au/water/coal-and-coal-seam-gas/regulation

Australian Department of Climate Change, Energy, the Environment and Water (DCCEEW). Resources and Energy Quarterly, September 2022. https://www.industry.gov.au/publications/resources-and-energy-quarterly-september-2022

Australian Department of Climate Change, Energy, the Environment and Water (DCCEEW). Resources and Energy Quarterly, December 2024b. https://www.industry.gov.au/publications/resources-and-energy-quarterly-december-2024

Baer, Hans. "The Nexus of the Coal Industry and the State in Australia: Historical Dimensions and Contemporary Channges." *Energy Policy* 99 (2016): 194–202.

Bamberry, Larissa. "Households and Citizenship in a Restructuring Region: The Latrobe Valley, Australia." *Labour & Industry* 25, no. 1 (2015): 23–37.

Barnes, Tom. "From Occupational Steering to Occupational Citizenship: Understanding Career Pathways After the Demise of Australian Automotive Manufacturing." In *Working through Planetary Breakdown: Labour, Skill and the Changing Climate*, edited by Chantel Carr and Jesse Adams Stein, Routledge. 2025.

Barnes, Tom, and Sally Weller. "Becoming Precarious? Precarious Work and Trajectories After Retrenchment." *Critical Sociology* 46, no. 4–5 (2020): 527–41.

Burrows, Scott. "Unemployment in the Illawarra Region: Understanding its Historical Trajectory and Legacy." *Illawarra Unity* 11, no. 1 (2012): 51–66.

Cha, J. Mijin. "A Just Transition for Whom? Politics, Contestation, and Social Identity in the Disruption of Coal in the Powder River Basin." *Energy Research & Social Science* 69 (2020): 101657.

Colvin, Rebecca. "Contextualizing Coal Communities for Australia's New Net Zero Authority." *Nature Energy* 8, no. 8 (2023): 776–8.

Colvin, Rebecca and Ewelina Przybyszewski. "Local Residents' Policy Preferences in an Energy Contested Region – The Upper Hunter, Australia." *Energy Policy* 162 (2022): 11276.

Connor, Linda. "Energy Futures, State Planning Policies and Coal Mine Contests in Rural New South Wales." *Energy Policy* 99 (2016): 233–41.

Dahlgren, Kari. "Blaming in the Boom and Bust: Greed Accusations in an Australian Coal Mining Town." *Cambridge Journal of Anthropology* 37, no. 2 (2019): 90–109.

Della Bosca, Hannah and Jo Gillespie. "The Coal Story: Generational Coal Mining Communities and Strategies of Energy Transition in Australia." *Energy Policy* 120 (2018): 734–40.

Deloitte Access Economics. *The Decentralisation of Work and the Illawarra: Practical Insights for Business*. Report Prepared for Business Illawarra, April 2021. https://www.deloitte.com/au/en/services/economics/perspectives/decentralisation-work-illawarra.html

Donaldson, Michael, Les Bursill, and Mary Jacobs. *A History of Aboriginal Illawarra, Volume 2: Colonisation*. Yowie Bay: Dharawal Publications, 2017.

Edwards, Gareth, Claire Hanmer, Susan Park, Robert MacNeil, Milena Bojovic, Jan Kucic-Riker, Dan Musil and Gemma Viney. *Towards a Just Transition From Coal in Australia*. Sydney: Sydney Environment Institute, 2022.

Ey, Melina and Meg Sherval. "Exploring the Minescape: Engaging with the Complexity of the Extractive Sector." *Area* 48, no. 2 (2016): 176–82.

Fairbrother, Peter. "When Politics Meets Economic Complexity: Doing Things Differently in the Gippsland Region, Australia." *Australasian Journal of Regional Studies* 23, no. 3 (2017): 400–20.

Fairbrother, Peter, Darryn Snell, Larissa Bamberry, Linda Condon, Scott McKenry, Tomi Winfree, Dean Stroud and Joanne Blake. *Jobs and Skills Transition for the Latrobe Valley. Phase 1: Benchmark Occupations and Skill Sets*. National Centre for Sustainability, Swinburne University of Technology and RMIT University Centre for Sustainable Organisations and Work, 2012.

Flanagan, Frances. "A Transition with Teeth: Reconfiguring the Concept of Just Transition to Recognise the Imperative of Fossil Fuel Cessation." In *Working Through Planetary Breakdown: Labour, Skill and the Changing Climate*, edited by Chantel Carr and Jesse Adams Stein, Routledge, 2025.

Fleming, David, Thomas Measham and Dusan Paredes. "Understanding the Resource Curse (or Blessing) Across National and Regional Scales: Theory, Empirical Challenges and an Application." *Agricultural and Resource Economics* 59 (2015): 624–39.

Geoscience Australia. "Coal," 2023. https://www.ga.gov.au/education/minerals-energy/australian-energy-facts/coal

Gibson, Katherine. "Hewers of Cake and Drawers of Tea: Women, Industrial Restructuring and Class Processes on the Coalfields of Central Queensland." *Rethinking Marxism* 5, no. 4 (1992): 29–56.

Gibson-Graham, J. K. *The End of Capitalism (As We Knew It)*. Minneapolis: University of Minnesota Press, 2006.

Goodman, James and David Worth. "The Minerals Boom and Australia's 'Resource Curse.'" *Journal of Australian Political Economy* 61 (2008): 201–19.

Head, Lesley, Carol Farbotko, Chris Gibson, Nicholas Gill and Gordon Waitt. "Zones of Fruction, Zones of Traction: The Connected Household in Climate Change and Sustainability Policy." *Australasian Journal of Environmental Management* 20, no. 4 (2013): 351–62.

Huber, Matthew T. *Climate Change is Class War*. London and New York: Verso, 2022.

Institute for Energy Economics and Financial Analysis (IEEFA). Australia's Coal Export Market: Shifting Trade Dynamics with Asia. July 30, 2024. https://ieefa.org/resources/australias-coal-export-market-shifting-trade-dynamics-asia

Hughes, Llewelyn, Thomas Longden and Yeliz Simsek. *Coal and Gas Futures in the Asia-Pacific: Analysis of the Impact of Climate Policy and Technological Pathways on the Energy Mix of China, India, Japan, South Korea and Indonesia*. Zero-Carbon Energy for the Asia Pacific ZCEAP Working Paper ZCWP3-23. ANU, 2023.

International Energy Agency (IEA). *Emissions Measurement and Data Collection for a Net Zero Steel Industry*. G7 Hiroshima Summit. IEA, 2023.

International Labor Organisation (ILO). *Guidelines for a Just Transition: Towards Environmentally Sustainable Economies and Societies for All*. International Labour Organisation, 2015.

MacNeil, Robert and Madeleine Beauman. "Understanding Resistance to Just Transition Ideas in Australian Coal Communities." *Environmental Innovation and Societal Transitions* 43 (2022): 118–26.

McDonald, Paula, Robyn Mayes and Barbara Pini. "Mining Work, Family and Community: A Spatially-Oriented Approach to the Impact of the Ravensthorpe Nickel Mine Closure in Remote Australia." *Journal of Industrial Relations* 54, no. 1 (2012): 22–40.

O'Brien, Martin, and Scott Burrows. "Assessing the Effectiveness of Regional Policy Responses to Mass Redundancies: The Case of the Illawarra Region, Australia." *Economic Papers* 38, no. 2 (2019): 144–55.

Parkinson, Giles. "Wind and Solar Smash Output Records as Grid Demand Hits New Highs in Midst of Heatwave." *Renew Economy*, December 16, 2024. https://reneweconomy.com.au/wind-and-solar-output-records-tumble-as-grid-demand-hits-new-highs-in-midst-of-heatwave/

Queensland Treasury, "State Budget 2024–25". *Tabled* 11 June 2024. https://www.treasury.qld.gov.au/resource/state-budget-2024-25/

Rainnie, Al and Darryn Snell. "The Australian Net Zero Economy Authority: Limits and Possibilities." In *Examining Net Zero: Creating Solutions for a Greener Society and Sustainable Economic Growth*, edited by Alex de Ruyter Alex, Steven McCabe and Beverley Neilsen, 25–34. Leeds: Emerald Publishing, 2025.

Sheldon, Peter, Junankar Raja and Pontello Anthony De Rosa. "The Ruhr or Appalachia? Deciding the Future of Australia's Coal Power Workers and Communities." *Report for CFMMEU Mining and Energy*. Industrial Relations Centre, UNSW, 2018.

Snell, Darryn. "'Just Transition?' Conceptual Challenges Meet Stark Reality in a 'Transitioning' Coal Region in Australia." *Globalizations* 15, no. 4 (2018): 550–64.

United Nations. *Paris Agreement to the United Nations Framework Convention on Climate Change*, December 12, 2015 T.I.A.S. No. 16–1104, 2015.

Vogl, Valentin. *Steel Beyond Coal: Socio-Technical Change and the Emergent Politics of Steel Decarbonisation*. PhD Thesis: Lund University, 2023.

Weller, Sally. "The Geographical Political Economy of Regional Transformation in the Latrobe Valley." *Australian Journal of Regional Studies* 23, no. 3 (2017): 382–99.

Weller, Sally. "Just Transition? Strategic Framing and the Challenges Facing Coal Dependent Communities." *Environment & Planning C: Politics and Space* 37, no. 2 (2019): 298–316.

Weller, Sally and Andrew Beer. "State Structures and the Limits of Agency: Governing the Transformation From Coal in Australia". *Regional Studies* 57, no. 8 (2022): 1415–27.'

8

INDUSTRIAL DEMOCRACY AND CLIMATE STRUGGLE

The Limits and Possibilities of Using Health and Safety Committees to Promote Industrial Democracy and Address Climate Change

Eugene Schofield-Georgeson, Michael Rawling and Brett Heino

Introduction

Australia's fossil fuel exports contribute significantly to high per capita emissions, placing it among the world's top emitters relative to both population and economy. In this context, capital and employers continue to be responsible for the majority of climate emissions, that remain largely unabated by government climate policy. Workers, meanwhile, will be amongst the groups most affected by climate change, and yet have to perform the labour that produces emissions. In a recent study, 75 percent of Australian workers indicated that they are already feeling the physical health impact of global warming and are demanding intervention.[1] Productive social relations, meanwhile, afford workers and trade unions unique power and motivation to act against climate collapse by resisting capital. Yet the Australian labour movement has been perilously weakened by 50 years of union decline and work fragmentation arising from neoliberal labour law and employer strategy. The ability of workers and unions to respond to environmental catastrophe therefore requires a double-movement that not only endows labour with the legal power to fight ecological crisis, but renews its strength to organise more broadly against capital. This chapter proposes that environmental labour strategy and trade union renewal share a common interest in renewing industrial democracy.

Industrial democracy involves employee control of work through bargaining, as well as participation in management decisions. Industrial democracy is "concerned with equalising power within the workplace."[2] A key legal mechanism enabling the practice of industrial democracy are Work Health and Safety Committees (HSCs). HSCs involve small groups of workers and management

DOI: 10.4324/9781003512240-11

representatives co-ordinating and participating in decisions affecting health, safety and employee wellbeing within firms. In this chapter we contend that through HSCs, workers may at once exercise both industrial power *and* environmental power. They may organise against their employers to ensure their own equal bargaining position, while also leading strategies to address climate mitigation (worker practices that support emissions reduction as a health and safety issue, for example) and adaptation (addressing heat-related work stress, such as that proposed by Humphrys in Chapter 9).[3]

The first author has written elsewhere of the efficacy of HSCs as a legal strategy for the labour movement to intervene in climate policy in an Australian industrial context.[4] This research identified HSCs as having strong legal powers (discussed below), enabling workers to enforce their own decision-making power against autocratic management decisions. While this case study is situated in the Australian context, its application is potentially global. HSCs are a recurring feature of health and safety regulation across the Global North where they are conventionally permitted a wide and discretionary ambit to operate as a mouthpiece for worker representatives within firms.[5] It is in this capacity that workers have leveraged HSCs for industrial democratic purposes. Little understood, however, is the incidence and impact of HSCs.

Accordingly, in the first section of this chapter we collate Australian statistical data regarding HSC use and efficacy in order to locate the potential for HSCs to take climate action and restore trade union power using the techniques of industrial democracy. The data is derived predominantly from a variety of peak union bodies. It generally reflects a progressive downturn in the use of industrial democratic mechanisms like HSCs, but they nevertheless persist in significant numbers, despite a comparable decline in trade union density. In response to these statistical findings, in the second section we propose new governmental policies and trade union organising strategies to enhance HSC efficacy, particularly in ways that allow workers to reduce emissions and adapt to unsafe climatic conditions.

We suggest that declining use of HSCs ought to be resolved, in part, through stronger legislative support for HSCs and broader work health and safety (WHS)/industrial reform. Reforming worker voice or access to industrial democracy in this way would enable multiple social improvements to WHS, while providing an avenue for worker-led climate mitigation and adaptation initiative and improved pay and conditions. Finally, in the third section we demonstrate how the particular synthesis of trade union and environmental struggle in HSCs can feature as part of a broader campaign that has the potential to rejuvenate and bolster union membership, presence and organisational capacity, each of which are crucial to ensuring the practice of industrial democracy.

Health and safety committees

In 2011, a federal Labor Government harmonised Australian work health and safety (WHS) legislation, ensuring uniformity across most aspects of the respective WHS Acts within sub-national states and territories. Every Australian WHS Act now affords workers the power to elect health and safety representatives (HSRs) and to form HSCs in order to assess, plan for, monitor and enforce workplace health and safety risks (e.g. *Work Health and Safety Act* (Cth), Part 5). In addition, workers may negotiate with employers to expand HSC powers to address a wide variety of environmental and other workplace issues.[6] Crucially, HSRs, most of whom are embedded within HSCs, have the power to issue an employer with a Provisional Improvement Notice (PIN) to rectify or ensure employer compliance with a directive developed by the HSC and/or HSR.[7] An employer who fails to comply with a PIN may be subject to a (maximum) $250,000 fine, issued by the relevant safety inspectorate.[8] The issuing of PINs by HSRs has been found to be 91–95 percent effective in producing employer compliance with health and safety directives.[9]

These legal powers mean that HSCs are a potentially potent democratic force within the otherwise autocratic and monopsonistic legal structure of medium and large employers.[10] Used in conformity with WHS legislation, these powers provide Australian workers with the collective ability to vocalise dissent and opposition to employer practices; but they can also require employers to change workplace practices backed by force of law. Despite the legal flourishing of these powers since 2011, however, longitudinal statistical data collected by trade unions and others show that HSCs are not being utilised to their full potential. They are in fact declining in use (see Table 8.1). [11]

As Table 8.1 shows, in 2022, the number of HSCs has halved since the peak of their take-up by workers in 2001.[12] These figures correlate with an overall decline in Australian trade union density from around 30 percent in 1998 to 12.5 percent in 2022.[13] Decreasing numbers of HSCs also reflect a small dip in worker perceptions of their effectiveness, shown over time in Table 8.2.[14]

TABLE 8.1 Percentage of Australian workplaces with HSCs over time

Year	*Workplaces with an HSC*	*Workplaces without an HSC*	*Data source*
1995	43%	30%	WIRS
2001	85%	11%	ACTU
2004	83%		VTHC
2005	81%	14%	ACTU
2021	49%		ACTU
2022	45%		ACTU

TABLE 8.2 Effectiveness of HSCs in implementing worker-led safety

Year	*Effective*	*Ineffective*	*Unsure*	*Source*
2001		43%	12%	ACTU
2004	63%			VTHC
2005		19%	13%	ACTU
2022	53%			ACTU

While less pronounced, the number of HSRs has also declined over time (Table 8.3),[15] along with recent perceptions of their effectiveness (Table 8.4).[16]

Perhaps a truer representation of the effectiveness of HSRs, however, is the rate at which they resolve health and safety issues with employers. On this score, HSRs have remained consistently effective (as shown in Table 8.5).[17] Particularly impressive is the comparison between rates of employer rectification of health and safety issues *without* HSRs in the workplace (45 percent in 2021); and employer rectification of health and safety issues when HSRs are present in the workplace (85 percent in 2021). Quite clearly, work, health and safety laws permit worker representatives to oversee health and safety decisions in the workplace and ensure their effective implementation. Increasingly, such action will involve climate adaptation policies, and we would argue there is no reason why they should not also extend to mitigation policy.

Arguably, were HSCs assessed on a similarly results-based metric as HSRs, they might score more highly than their current "effectiveness" rating, based on perceptions. In this respect, the disparity between HSR and HSC effectiveness may also reflect training or comprehension of the ability for HSRs and HSCs to co-ordinate their activities. Data on HSR training consistently shows that only around one quarter to a third of HSRs have undergone training in

TABLE 8.3 Percentage of Australian workplaces with HSRs over time

Year	*Percentage*	*Source*
1995	66%	WIRS
2019	53%	ACTU
2021	63%	ACTU
2022	56%	ACTU

TABLE 8.4 Effectiveness of HSRs in implementing worker-led safety demands

Year	*Effective*	*Source*
2021	85%	ACTU
2022	78%	ACTU

TABLE 8.5 Problems identified by HSRs, resolved by employers

Year	*Problems identified by HSRs resolved by employers*	*Source*
2001	77%	ACTU
2005	79%	ACTU
2021	85%	ACTU
2021	45% (compliance without HSRs)	ACTU
2022	78%	ACTU
2022	50% (compliance without HSRs)	ACTU

order to understand their strong legal powers and that they can be effectively co-ordinated with the activities of HSCs.[18] As the next section of this chapter shows, co-ordination of HSR and HSC activities may strengthen their roles, by collectivising or sharing some of the problems experienced by HSRs on an individual basis (e.g. negative pressure, intimidation, harassment and bullying). Such strategies would also enhance the ability of HSCs to make and implement climate policy in the workplace.

Accordingly, while the legal power for workers to take effective environmental action against capital may already exist,[19] the potential for such action is moderately declining and is by no means well utilised. This moderate decline in worker climate action through HSCs correlates with declining trade union density, and poor levels of HSR training to coordinate their activities with HSCs. As the later parts of this chapter make clear, however, government policy (discussed in the next section) and trade union organising strategy (discussed in the final section) to rebuild membership density, hold the potential to reinvigorate industrial democracy and thereby allow workers to take climate action against employers.

HSCs, industrial democracy and government policy

Industrial democracy measures such as worker representation and consultation are the means by which the voices of workers can be heard and acted upon for the benefit of workers[20] and the environment. HSRs and HSCs are central to debates about *existing* mechanisms that promote industrial democracy in Australia, given HSRs and HSCs are a form of worker voice within enterprises. There are both possibilities for, and limits to, using HSRs and HSCs to promote industrial democracy and worker-led addressing of climate issues. At a practical level, such issues might include decisions about auditing the workplace for environmental efficiency, investing in solar panels, switching to electric vehicles, planting trees, conserving water, using less plastic and non-renewable materials, implementing less carbon-intensive production processes and supply chains. In the main, because HSCs deal with the WHS issues that are often central to broader issues over the control of work and production, they show significant potential. However, the labour movement could encourage Parliaments to legislate in order to increase the powers of worker representatives on HSCs. In particular, we suggest that declining use of HSCs ought to be resolved, in part, through stronger legislative support for HSCs and broader labour law reform.

In terms of the potential of HSCs: their decisions can bind employers and can be enforced by HSRs; they can be used to achieve worker and union monitoring of WHS at a workplace; they can encourage union membership and are not restricted to unionised workplaces; and workers in Australia have the legislated right to be paid wages to perform WHS work.[21] In addition, once established, HSRs and HSCs can form an ongoing worker voice within the corporation.[22] Once formed, as we have seen in the first part of this chapter, HSRs and HSCs can be quite effective in promoting the health and safety of workers. Although there is some complexity with WHS issues, they can be a good motive to organise and recruit union members.[23] HSRs and HSCs can also be useful vehicles to build awareness of the role of unions.[24]

Although they address specific WHS purposes and these can be presented as separate, "technical" issues, Quinlan[25] highlighted some time ago, the interrelatedness of OHS and other issues as an "inseparable element of the industrial relations mosaic." This in turn, reveals the potential for HSCs to address climate issues including the pursuit of work practices that can reduce emissions reduction and combatting climate-related work hazards. WHS might be used as an integral part of general representational and organising work.[26] Worker representation might for example extend to matters such as work organisation and working time as these can adversely affect WHS.[27] Such an approach is backed by research evidence that clearly establishes a link between WHS and other workplace issues.[28] WHS has considerable significance to broader industrial relations because it is a part of the experience of workers and interacts with their other work experiences. It is an issue that can raise broader issues of the control over work and production.[29] Furthermore, features of work and production which give rise to industrial relations problems such as autocratic management, lack of worker autonomy, contract and casual labour, rostering and incentive-based pay systems are also closely associated with risks to workers' health and safety.[30] In some industries in which piece rates are used, such as in the road transport, home-based clothing manufacturing and horticulture industries, the link with broader industrial relations issues is plainly evident. Here low pay rates contribute to unsafe working conditions.[31]

HSRs and HSCs present an opportunity to progress workplace democracy on these broader concerns through the prism of WHS issues. But questions remain about the extent of their mandate to cover a broader variety of interrelated WHS and industrial issues. In particular, the WHS regime, which HSRs and HSCs fall under, mostly or wholly exclude consideration of the economic pressures that lead to hazardous working conditions.[32] Thus, the conventional separation of WHS from the regulation of pay and conditions is an ongoing hurdle to the effectiveness of WHS instruments in propelling forward effective forms of workplace democracy.

Thus the considerable limits to the current HSR and HSC structure include that they are established under a somewhat siloed regime of WHS law, separate to the major Australian industrial legislation (the *Fair Work Act 2009*).[33] Meanwhile, HSR and HSC mechanisms may not institute effective worker voice through a wide range of workplaces because they rely on collectivities of workers, ideally sourced from medium to large-sized enterprises.[34] There also may be a concentration of HSCs and HSRs in industrial sectors given heightened WHS risks in those sectors. The scope of HSR and HSC powers remain uncertain and in practice may exclude interrelated industrial relations (like the effect of pay rates on safety) and climate issues (such as employer emissions). And, at times, managers may attempt to send matters to the HSCs to stall progress on WHS issues. These problems raise questions about the current effectiveness of some HSCs in certain workplaces, especially those without a strong union presence, as identified through the statistical analysis above. Such issues also point to similar solutions involving legislative reform and union strategy.

Perhaps the key shortcoming of the legal framework is that, in all Australian jurisdictions, there is no mandatory requirement for employers to establish HSRs and HSCs unless requested to do so.[35] There is no universal access to this form of worker representation. Instead, the onus is placed on workers rather than employers to initiate their establishment.[36] Therefore, the rise in employer power over previous decades has adversely affected the ability of HSRs and HSCs to take full advantage of their legislated rights.[37] As a consequence, the potential for HSCs and HSRs to further industrial democracy (and address climate change) across the economy has not been realised.

This is not just a theoretical quibble. Take the recent example of attempts begun in 2020 by workers and the Transport Workers' Union (TWU) to get HSRs elected at Deliveroo to represent food deliverers in Sydney. This involved almost year-long negotiations between the TWU and Deliveroo over the relevant groups of workers to be represented by the HSRs, as well as two industrial commission decisions. HSRs were eventually elected, but well over two and half years after the process to initiate HSRs had begun, the issue was not fully resolved because of the ability of Deliveroo to raise technical issues with the process.[38] One lesson to be drawn from this example is that for work groups to be negotiated and HSRs elected, the support of a committed union is required. Large and well-resourced companies can drag out negotiations and contest legal issues. In these sorts of cases HSRs face significant challenges and can spend much of their time arguing about their functions rather than engaging in WHS work.[39]

For HSRs and HSCs to be effective requires their implementation in the workplace. The most complete implementation is frequently concentrated in larger workplaces where there is a unionised workforce.[40] This further demonstrates that for HSCs and HSRs to be most effective requires strong union workplace presence.[41] Yet such a union presence has not been maintained across the Australian economy in recent decades and this has been a major hurdle for effective HSRs and HSCs.[42]

The implication here is that HSCs have significant *potential* to boost industrial democracy but *there are preconditions for their success*.[43] These preconditions include appropriate state regulation and support from unions.[44] Support from unions is especially important for HSRs to communicate effectively with their worker constituencies and to build trust with those constituencies.[45] This is one of the reasons why trade union WHS training is important. As HSRs develop a better understanding of WHS, this may not only bring them into conflict with employers but can distance them from the views of workers they represent.[46] Studies have shown that trained HSRs "make for better participative practices."[47] However, this demonstrates that the current system generates significant resource burdens on unions and workers.[48] Ultimately, for HSCs and HSRs to provide an effective voice for labour (including on the point of resisting and changing climate-impacting activities of capital), further legal reforms are necessary. Such reforms would – most importantly – expand the issues on which employers are required to consult workers, as well as obliging *all* employers to establish consultative arrangements.[49]

Trade union strategy and environmental partnership

Whatever the fecundity of the relationship between HSCs/HSRs and industrial democracy, the current structure of the law imposes substantial restrictions on this articulation. In this section, therefore, we seek to frame the earlier sections by going beyond the specific relationship between HSCs/HSRs and industrial democracy to explore a broader problematic – the degree to which trade union involvement in environmental issues fertilises union renewal and industrial democracy. This analysis identifies vectors of mobilisation that might augment the preconditions for more meaningful industrial democracy identified previously, and creates a richer picture of the industrial democracy/environmental action relationship of which HSCs/HSRs can be part.

While not ignoring the challenges facing unions in this space (including the issue of limited resources), we nevertheless suggest that there are a number of interrelated advantages to unions in taking up the struggle over environmental issues. These include:

- Opportunities to grow/mobilise union membership;
- Coalition building and, in turn, increased political influence;
- Augmenting the way that unions struggle within and against capitalism.

At the most basic level, campaigning on environmental issues might offer opportunities to increase and/or better mobilise union membership. Although studies demonstrating an explicit link between environmental struggles and union growth are lacking, it is logical that such a relationship should exist. Numerous studies have demonstrated that concern about environmental issues such as climate change is rapidly increasing, particularly amongst younger generations (the very group of workers unions have historically had the most difficulty organising).[50] At the same time, it has been observed that union members are more pro-environmentally inclined than non-members, even in industries such as transport and manufacturing where the effect of climate change mitigation is most pronounced.[51] Although necessarily speculative, there is perhaps a virtuous circle to be found here; other things being equal, a union campaigning on environmental issues is more likely to be attractive to environmentally-conscious workers (particularly young workers). Even if there is considerable ambiguity as to whether or not such a focus leads to increased raw numbers of members,[52] an "environmental union agenda may serve to attract and engage new activists."[53] This could contribute to the cause of union renewal through infusing a new energy into both union memberships and networks of delegates and officials.

Beyond simply contributing to increased numbers of members, union involvement in environmental matters has the potential to greatly augment union strength through coalition-building. If we take a broad view of environmental issues i.e. beyond a singular focus on climate change, it is clear that unions have actively mobilised around them for decades.[54] Australian unions are well-represented in this space. One of the leading historical examples in this context is the activities of the New South Wales branch of the Builders Labourers Federation (BLF), which pioneered the use of so-called "Green Bans" to protect both natural and built habitats from over-development. In relation to the struggle to protect Sydney's Kelly's Bush, BLF official Jack Mundey offered an expansive view of the role of the union in community affairs:

> I summed up the discussion by saying what was the use of winning higher wages and better conditions if we lived in cities devoid of parks and denuded of trees? Our cities had to be for people, not for corporations to plunder and destroy. Kelly's Bush wasn't just for its neighbours, it should be public land and used by everybody who wanted to use it.[55]

By assuming responsibility for the health of both natural and built environments, the BLF found its position in public affairs radically elevated. The union provided the foundation of a cross-community alliance that united construction workers, community activists and progressive professionals.[56] Indeed, the members of the "Battlers for Kelly's Bush" were primarily women who had voted for the conservative Liberal Party their entire lives.[57] Environmentally-oriented unionism, therefore, proved the pivot point that allowed the BLF to declare and enact radical political-economic concepts including direct action, everyday democracy, quality of life and worker control, concepts that allowed it to formulate its actions as those of "the people."[58]

The utility of environmental issues acting as a glue binding together unions and other community groups has been demonstrated time and again in other political-economic contexts as well. In the USA, so-called "Blue-Green" alliances mobilised in the 1960s and 1970s around issues "such as air and water protection, in addition to issues of workplace health and safety," and proved central in the passage of key pieces of environmental legislation, including the *Clean Air Act*, *Clean Water Act* and the *Toxics Substances Control Act*.[59] Across Europe, struggles against water privatisation have combined unions and community groups into a powerful front against a particularly sinister form of neoliberal restructuring.[60] Meanwhile, across Asia and Africa, union-led coalitions have tackled issues as diverse as reforestation, clean water and hazardous mining.[61] These struggles obviously go well beyond the specific issue of HSCs, industrial democracy and environmental action that has been the focus of this chapter, but it demonstrates that the latter does not exist in a vacuum. Rather, the specific articulation between industrial democracy and environmental action potentially crystallised in HSCs can draw upon a broader tradition of action at the interface of organised labour, the community and the environment, a tradition that can certainly serve a validating function.

Beyond the generation of general goodwill (which is of course a very valuable resource for any union), we would note one chief feature of green coalition building that directly contributes to union renewal and the possibilities for deeper industrial democracy. Given the fact that ecological crises and repair are now explicitly political concerns, mobilising around them can open additional avenues through which unions can influence outcomes at the political level. This is particularly so regarding input into the legislative sphere. In many of the examples of coalition-building cited above, union action either heavily-influenced, or directly contributed to, the design of new legislation. This capacity can create opportunities whereby environmental legislation can be synergised with labour-friendly provisions, encapsulated in the notion of a "just transition" (see Frances Flanagan in Chapter 6 of this collection).[62] Indeed, two recent examples attest strongly to this opening. In 2018, the central Spanish government passed a deal with unions that will see all of Spain's private coalmining pits closed. The deal "mixes early retirement schemes for miners over 48, with environmental restoration work in pit communities and re-skilling schemes for cutting-edge green industries."[63] Across the Atlantic, in 2023 a landmark piece of legislation, the *Build Public Renewables Act*, requires the publicly owned New York Power Authority (NYPA) to take wide-ranging steps to meet the state's goal of 70 percent renewable electricity generation by 2030. Union involvement, which was facilitated by the New York chapter of the Democratic Socialists of America and Public Power New York, has helped implant powerful labour provisions, including mandatory collective bargaining agreements for all employees, contractors and subcontractors involved on NYPA renewable projects, and a twenty-five million dollar just transition fund.[64] In both cases, union mobilisation around environmental matters translated into meaningful pro-labour terms embedded in green legislation, terms that represented an objective advance over the broader labour law regime.

Finally, union assumption of responsibility for environmental issues offers a powerful conceptual restatement of the role of trade unions within and against modern capitalism. The reality of union decline since the 1970s has of course been established beyond any doubt. Deindustrialisation in the developed world, the advent of widespread employment precarity and the fissuring of work relations into a myriad of legal and illegal forms[65] has rendered unviable any simplistic notion of "bread and butter" unionism focused solely on wages and working conditions. Indeed, the nature of value creation and surplus value extraction today spans a whole host of social domains, no longer being exclusively concentrated in a spatially-delimited workplace.[66] Modern capitalism, in other words, organises and mobilises the whole of society to produce profit.[67] Unions struggling over environmental issues, of which HSCs/HSRs are but one example, is important recognition of this reality and powerfully reaffirms the importance of organised labour.[68] It is perhaps the most powerful new force that could inform a "social movement unionism" for the twenty-first century. Social movement unionism combines "[d]emocracy, solidarity, militancy, internationalism, independence, workplace organisation and activism."[69] Hampton argues that ecological issues can be readily absorbed within this framework, a contention we whole-heartedly agree with.[70] Given the fact that contemporary ecological breakdown by nature cannot be restricted to specific localities and jurisdictions, such ecologically-inspired social movement unionism must perforce place a premium of responsiveness to the membership, a willingness to lead non-union workers and groups, and global working-class solidarity. Although the obstacles in the path of that brand of unionism are formidable, the kind of industrial democracy which it could constitute is worth fighting for.

Conclusion

Industrial democracy is a powerful force for social and environmental change that operates by giving voice to workers in the productive sphere. The legal mechanisms we have identified – HSCs and HSRs – provide a sophisticated and already-available toolkit by which workers can practise social democracy and make change. However, as this chapter has shown by reference to trade union statistics and an accompanying literature, the potential for workers to harness industrial democracy for environmental purposes is dwindling. Decline is associated with a range of factors, chiefly concerning a corresponding decline in trade union power and a lack of worker training in how to mobilise HSCs.

In response, the authors have pointed to two sets of interlocking solutions bringing together legislative action by the state and trade union organising strategy. The problem demands a state policy response that: expands the issues on which employers are required to consult workers through HSCs and HSRs; and that requires *all* employers to establish consultative arrangements such as HSCs and HSRs. In tandem, there is an opportunity for trade unions to partner more closely with environmentalist organisations. It is through such coalitions and alliances that unions can seize opportunities to regrow and mobilise their membership, thereby increasing their political influence. Augmenting the relationships established by trade unions in the productive sphere is therefore crucial to the survival trade unions and their ability to fight climate collapse.

Notes

1 Denham and Rickards, "Climate Impacts at Work," 9 and 73–76.
2 "ACTU Policy on Industrial Democracy," cited in Ford and Plowman, *Australian Unions*, 503.
3 See Chapter 9 of this volume: Humphrys, "Hot under the PPE Collar."
4 Schofield-Georgeson, "Legal Obstacles," 297. This contribution draws on Markey and McIvor, "Environmental Bargaining in Australia," and Hyman, *Understanding European Trade Unionism.*
5 Walters and Nichols, *Worker Representation*; Walters and Nichols, *Workplace Health and Safety*, "The Australian Framework," 31.
6 *WHS Act* (Cth), s76(1), s77(c); *NSW Trains* FCA 883; Schofield-Georgeson, "Legal Obstacles," 305.
7 *WHS Act*, s90.
8 *WHS Act*, s99.
9 Australian Council of Trade Unions (ACTU), *A Report on the 2004 National Survey*, cited in Johnstone, "Institutional Arrangements," 42; ACTU, *A Report on the 2001 National Survey*, 10.
10 Nyberg and Murray, "Corporate Populism"; Manning, *Monopsony in Motion*.
11 Sources: Morehead et al., *Changes at Work: The 1995 Australian Workplace Industrial Relations Survey* (WIRS), 20; ACTU, *A Report on the 2001 National Survey*, 8; ACTU, *A Report on the 2004 National Survey*, cited in Johnstone, 39; Victorian Trades Hall Council (VTHC), *The View From the Front Line*, 8; ACTU, *Work Shouldn't Hurt*, 41; ACTU, *Work Shouldn't Hurt: Survey Report*, 36.
12 Data has been compiled from various sources as indicated in "Data Source" column. Statistics on HSCs are not centrally collected and there is no publicly available reporting for evident data gaps.
13 Australian Bureau of Statistics (ABS), "Trade Union Membership."
14 ACTU, *A Report on the 2001 National Survey*, 8; VTHC, *The View From the Front Line*, 8; ACTU, *A Report on the 2004 National Survey*, cited in Johnstone, 39; ACTU, *Work Shouldn't Hurt* (2022), 36. These sources provide a rich, albeit patchy (Johnstone, "Institutional Arrangements," 39) set of longitudinal data on the operation of HSCs.

15 Morehead et al., *Changes at Work*, 20; ACTU, *Work Shouldn't Hurt* (2019), 11; ACTU, *A Report on the 2001 National Survey*, 36; The difference between the 2021 and 2022 rates of HSRs is due to a note inserted into the 2022 survey explaining that the role of an HSR is as "a worker who has been elected/selected by a group of workers to represent them on health and safety matters." What this may indicate is that around six percent of HSRs in the 2021 survey may have been employer appointments, rather than elected by co-workers.
16 ACTU, *Work Shouldn't Hurt* (2021), 43; ACTU, *Work Shouldn't Hurt* (2022), 37. Both studies involved a sample size of around 1,500 participants.
17 ACTU, *A Report on the 2001 National Survey*, 8; ACTU, *A Report on the 2004 National Survey*, cited in Johnstone, 41; ACTU, *Work Shouldn't Hurt* (2021), 43; ACTU, *Work Shouldn't Hurt* (2022), 36–37.
18 ACTU, *A Report on the 2001 National Survey*, 8; ACTU, *A Report on the 2004 National Survey*, cited in Johnstone, 39; ACTU, *Work Shouldn't Hurt* (2022), 35.
19 Schofield-Georgeson, "Legal Obstacles," 307–18.
20 Walters and Nichols, "Worker Representation," 144.
21 Schofield-Georgeson, "Legal Obstacles," 311–12.
22 This can be compared to the periodic worker voice that is conveyed through collective bargaining: at least in Australia, collective bargaining mainly provides workers with the ability to be heard when bargaining and taking industrial action for a new collective agreement. Once the agreement has been agreed and registered there is not the same ability to exercise voice for a period of years until the agreement expires and a new round of bargaining commences.
23 Loudon and Walters, "Trade Union Strategies," 192, 195.
24 Walters and Nichols, *Worker Representation*, 164.
25 Quinlan, "The Industrial Relations of Occupational Health and Safety," 144.
26 James, "Worker Representation and Health and Safety," 214.
27 James, "Worker Representation and Health and Safety," 213.
28 Loudon and Walters, "Trade Union Strategies," 190.
29 Loudon and Walters, "Trade Union Strategies," 190.
30 Loudon and Walters, "Trade Union Strategies," 146.
31 Bluff et al., "Regulating Health and Safety in Work," 109.
32 Bluff et al., "Regulating Health and Safety in Work," 108.
33 Schofield-Georgeson, "Legal Obstacles," 306.
34 Schofield-Georgeson, "Legal Obstacles," 311.
35 Stewart et al., *Creighton and Stewart's Labour Law*, 565; James, "Worker Representation and Health and Safety," 204.
36 James, "Worker Representation and Health and Safety," 208.
37 James, "Worker Representation and Health and Safety," 206; see also the first section of this Chapter.
38 Bluff et al., "Regulating Health and Safety in Work," 104.
39 Bluff et al., "Regulating Health and Safety in Work," 106.
40 Loudon and Walters, "Trade Union Strategies," 183.
41 Loudon and Walters, "Trade Union Strategies," 180; Walters and Nichols, *Worker Representation*, 148.
42 Loudon and Walters, "Trade Union Strategies," 190.
43 Walters and Nichols, *Worker Representation*, 144–45.
44 Walters and Nichols, *Worker Representation*, 139.
45 Walters and Nichols, *Worker Representation*.
46 Walters and Nichols, *Worker Representation*, 140.
47 Walters and Nichols, *Worker Representation*, 141.
48 Schofield-Georgeson "Legal Obstacles," 297.
49 Walters and Nichols, *Worker Representation*, 151–52.
50 See for example Peetz et al., "Ageing Australian Unions."

51 Ringqvist, "Union Membership."
52 See, for example, Ringqvist, "Union Membership," 677; Farnhill, "Union Renewal and Workplace Greening," 716.
53 Ringqvist, "Union Membership," 677.
54 On this point, see also, Snell, Fairbrother, and Hart, "Blue-green Alliances."
55 Mundey, *Green Bans and Beyond.*
56 Iveson, "Building a City for 'The People.'"
57 "Kelly's Bush Park," *State Heritage Inventory.*
58 Iveson, "Building a City for 'The People,'" 998–1003.
59 Contorno, "Turtles and Teamsters Revival?," 120.
60 Bieler, *Fighting for Water.*
61 Bureau for Workers' Activities, *Trade Union Actions.*
62 Flanagan, "A Transition with Teeth."
63 Nelsen, "Spain to Close Most Coalmines."
64 Goldmansour, "How New York's Democratic Socialists Brought Unions Around."
65 Dukes and Streeck, *Democracy at Work.*
66 Mason, *PostCapitalism*, 139.
67 Tronti, "Factory and Society."
68 Huber, *Climate Change as Class War.*
69 Hampton, *Workers and Trade Unions*, 36.
70 Hampton, *Workers and Trade Unions*, 36–37.

Bibliography

ABS. "Trade Union Membership: August 2022." Released 14 December 2022. https://www.abs.gov.au/statistics/labour/earnings-and-working-conditions/trade-union-membership/latest-release

ACTU. *A Report on the 2001 National Survey of Health and Safety Representatives.* Melbourne, 2002.

ACTU. *Work Shouldn't Hurt: A Survey on the State of Work Health and Safety in Australia.* Melbourne, 2019.

ACTU. *Work Shouldn't Hurt: Survey Report 2022.* Melbourne, 2022a.

ACTU. *Work Shouldn't Hurt: The State of Work Health and Safety in Australia 2021.* Melbourne, 2022b.

Bieler, Andreas. *Fighting for Water: Resisting Privatization in Europe.* London: Bloomsbury, 2021.

Bluff, Elizabeth., Richard Johnstone and Michael Quinlan. "Regulating Health and Safety in Work for Digital Labor Platforms in Australia: The Example of Food Deliverers." *Journal of Work Health and Safety Regulation* 1, no.1 (2023): 92–116.

Bureau for Workers' Activities, *Trade Union Actions to Promote Environmentally Sustainable Development.* Geneva: International Labour Organization, 1999.

Contorno, Lauren. "Turtles and Teamsters Revival? Analyzing Labor Unions' Environmental Discourse from the 2014 People's Climate March." *Interface: A Journal For and About Social Movements* 10, no. 1–2 (2018): 117–48.

Denham, Todd and Lauren Rickards. *Climate Impacts at Work: Supporting a Climate Ready Workforce.* Research Report, September 2022.

Dukes, Ruth and Wolfgang Streeck. *Democracy at Work: Contract, Status and Post-Industrial Justice.* Cambridge: Polity Press, 2023.

Farnhill, Tom. "Union Renewal and Workplace Greening: Three Case Studies." *British Journal of Industrial Relations* 56, no. 4 (2018): 716–43.

Ford, Bill and David Plowman, eds. *Australian Unions: An Industrial Relations Perspective*. South Melbourne: MacMillan, 1983.

GoldMansour, Luca. "How New York's Democratic Socialists Brought Unions Around to Public Renewables." *The American Prospect*, 19 June 2023. https://prospect.org/environment/2023-06-19-new-york-democratic-socialists-unions-public-renewables/

Hampton, Paul. *Workers and Trade Unions for Climate Solidarity: Tackling Climate Change in a Neoliberal World*. London: Routledge, 2015.

Huber, Matthew T. *Climate Change as Class War: Building Socialism on a Warming Planet*. London: Verso, 2022.

Hyman, Richard. *Understanding European Trade Unionism: Between Market, Class and Society*. London: Sage, 2001.

Iveson, Kurt. "Building a City For 'The People': The Politics of Alliance-Building in the Sydney Green Ban Movement." *Antipode*, 46, no. 4 (2014): 992–1013.

James, Phil. "Worker Representation and Health and Safety: Reflections on the Past, Present and Future." In *Workplace Health and Safety: International Perspectives on Worker Representation*, edited by David Walters and Theo Nichols, 201–15. London: Palgrave Macmillan, 2009.

Johnstone, Richard. "Institutional Arrangements on Health and Safety Representation in Australia." In *Workplace Health and Safety: International Perspectives on Worker Representation*, edited by David Walters and Theo Nichols, 38–41. London: Palgrave MacMillan, 2016.

Johnstone, Richard. "The Australian Framework for Worker Participation in Occupational Health and Safety." In *Workplace Health and Safety: International Perspectives on Worker Representation*, edited by David Walters and Theo Nichols, 31–49. London: Palgrave MacMillan, 2009.

Loudon, Rebecca, and David Walters. "Trade Union Strategies to Support Representation on Health and Safety in Australia and the UK: Integration or Isolation?" In *Workplace Health and Safety: International Perspectives on Worker Representation*, edited by David Walters and Theo Nichols, 177–200. London: Palgrave Macmillan, 2009.

"Kelly's Bush Park," *State Heritage Inventory*. https://www.hms.heritage.nsw.gov.au/App/Item/ViewItem?itemId=5045753

Manning, Alan. *Monopsony in Motion: Imperfect Competition in Labor Markets*. Princeton: Princeton University Press, 2003.

Markey, Ray, and Joe McIvor. "Environmental Bargaining in Australia." *Journal of Industrial Relations* 61, no. 1 (2019): 79–104.

Mason, Paul. *PostCapitalism: A Guide to Our Future*. London: Allen Lane, 2015.

Morehead, Alison et al. *Changes at Work: The 1995 Australian Workplace Industrial Relations Survey*. Australian Government; Department of Workplace Relations & Small Business, 1997.

Mundey, Jack. *Green Bans and Beyond*. London and Sydney: Angus and Robertson, 1981.

Nelsen, Arthur. "Spain to Close Most Coalmines in €250m Transition Deal." *The Guardian*, 27 October 2018. https://www.theguardian.com/environment/2018/oct/26/spain-to-close-most-coal-mines-after-striking-250m-deal

NSW Trains v Australian Rail, Tram and Bus Industry Union ("NSW Trains") FCA 883 [44] – [45].

Nyberg, Daniel and John Murray. "Corporate Populism: How Corporations Construct and Represent 'the People' in Political Contestations." *Journal of Business Research* 162, (2023): 113879.

Peetz, David, Robin Price and Janis Bailey. "Ageing Australian Unions and the 'Youth Problem'." In *Young Workers and Trade Unions: A Global View*, edited by Andy Hodder and Lefteris Kretsos, 54–70. Basingstoke: Palgrave Macmillan, 2015.

Quinlan, Michael. "The Industrial Relations of Occupational Health and Safety." In *Work and Health: The Origins, Management and Regulation of Occupational Illness*, edited by Michael Quinlan, 140–70. Melbourne: Macmillan, 1993.

Ringqvist, Josef. "Union Membership and the Willingness to Prioritize Environmental Protection Above Growth and Jobs: A Multi-level Analysis Covering 22 European Countries." *British Journal of Industrial Relations* 60, no. 3 (2022): 662–82.

Schofield-Georgeson, Eugene. "Legal Obstacles and Possibilities for Environmental Bargaining in Australia." *Journal of Industrial Relations* 65, no. 3 (2023): 297–320.

Snell, Darryn, Peter Fairbrother and Alison Hart. "'Blue-green Alliances': Union Capacities and Prospects as Environmental Actors." (Conference Paper). *Australian Sociological Association Conference 2009 Annual Conference*, 2009. https://www.academia.edu/27945983/Blue_Green_Alliances_union_capacities_and_prospects_as_environmental_actors

Stewart, Andrew, et al. *Creighton and Stewart's Labour Law* (6th ed.). Annandale, NSW: Federation Press, 2016.

Tronti, Mario. "Factory and Society." *Operaismo in English*, 13 June 2013. https://operaismoinenglish.wordpress.com/2013/06/13/factory-and-society/

Walters, David and Theo Nichols. *Worker Representation and Workplace Health and Safety*. New York: Palgrave Macmillan, 2007.

Walters, David and Theo Nichols. *Workplace Health and Safety: International Perspectives on Worker Representation*. Basingstoke: Palgrave Macmillan, 2009.

Work Health and Safety Act 2011 (Cth).

9

HOT UNDER THE PPE COLLAR

Occupational High Heat During the Covid-19 Pandemic

Elizabeth Humphrys

Introduction

Rising temperatures, more prevalent and longer heatwaves, and increasing extreme weather, all impact workers' health and wellbeing.[1] Occupational heat stress is the most significant problem for labour related to climate change, and leads to increased injuries, illness and excess deaths.[2] These impacts are set to worsen in coming decades. Australia is a useful indicator for many countries in the Northern Hemisphere because projections place most of Australia's land mass in the moderate to high-risk category for occupational heat stress, and some parts of the north-west of the country at extreme risk.[3] In this sense, what happens in Australia now may herald conditions that may develop over slightly longer timeframes in other parts of the world.

In response to the need for more detailed qualitative research into the work health and safety (WHS) experience of workers as a result of climate change impacts, an 18-month project was conducted with the United Workers Union (UWU) in Australia. UWU is one of the largest and most diverse blue-collar unions in Australia, with approximately 150,000 members across 65 industries. Many members work in heat-exposed environments, in both indoor and outdoor settings, including early childhood education and care, home care, cleaning, warehousing, manufacturing, utilities, call centres, security, on farms, in restaurants, and as teacher's aides. The project commenced just as the Covid-19 pandemic began, which provided a window into how the two health and safety challenges – of occupational heat and the pandemic – were managed concurrently in a wide variety of workplaces. Of course, no WHS or climate change related crisis exists in isolation, these are always overlapping – high temperatures with bushfires,

DOI: 10.4324/9781003512240-12

flooding with the crisis of inequality, climate change with other environmental problems, or all these things at once – and these must be tackled concurrently.[4]

This chapter commences with an overview of research on the impact of Covid-19 related measures on the experience of occupational heat, to establish the additional and serious WHS heat risks workers faced during the pandemic. The second section details the methodology for the research and overviews what UWU members advised about their experiences. It explores the ways in which changes to working arrangements because of the pandemic exacerbated negative experiences of occupational heat, focusing on the impact of new forms of personal protective equipment (PPE), changes to working schedules and processes, and the imposition of new duties. The final section explores the implications of the research. It argues that the tension between pandemic measures and WHS protocols in relation to occupational heat largely resulted in the former trumping the latter, with employers prioritising productivity and cost concerns even when health risks were clear and mitigating action was possible. This indicates worker safety can be compromised when complex and multiple safety issues, especially new risks, intersect.

Background and methodology

Just prior to commencing interviews with UWU members for a broader project on occupational heat and climate change, the Covid-19 pandemic was declared on 11 March 2020 by the World Health Organisation. Less than two weeks later Australia closed its national borders and declared its first of many national and city-wide lockdowns in pursuit of a successful "zero Covid-19" strategy, until vaccines were available. This approach was based on almost immediate external and internal border closures on 20 March 2020, mandatory testing and isolation periods, use of quarantine facilities for limited in-country arrivals, extensive tracing of positive results and their contacts, and generalised restrictions on movement, physical contact, and time outside of the home. Over time, many states within Australia closed their borders to each other, with permission needed to move across them, and various cities and local government areas also prevented residents from travelling other than for essential health and work reasons.[5] The new government infection control protocols and mandates were enforced via severe financial penalties and active policing, which resulted in significant changes in workplaces and wider society. These zero Covid-19 strategies were relaxed in mid- to late 2021, after several outbreaks nationally. The strategy resulted in the country being effectively virus-free in the research period, from September 2020 to March 2021, and without major labour shortages contributable to Covid-19 related illness and isolation. However, many workers involved in this study faced similar pressures as in other countries, vis a vis heat and the impacts of PPE on working conditions.

For much of the years 2020 and 2021 Australia's two most populous states, New South Wales and Victoria – with a combined population of almost 15 million, around 60 percent of the national population – ordered that anyone who was reasonably able to work from home was mandated to do so. Those working in indispensable roles, often called "essential workers," remained working at their usual workplace, but had to comply with new rules and practices. These workers evidently included nurses and doctors, but also some early childhood education and care workers, cleaners, delivery drivers, warehouse workers, laundry workers, manufacturing workers, and many involved in food transport and delivery. Other changes included new PPE and uniforms, as well as social distancing and mandatory home isolation for 14 days if infected or exposed to someone who tested positive. While the low infection numbers in Australia were quite different to many other countries, the nature of restrictions and their impact on WHS meant workers often had experiences comparable to other locations.

In these circumstances Australia was reliant on its essential workers to keep running, and such people were lauded in the media and by government for their efforts in difficult circumstances. UWU members were under extraordinary pressure in roles as cleaners, in supermarket supply chains, in the agricultural sector, providing transport and logistics, and in security roles in mandatory pandemic hotel quarantine facilities for those arriving in the country. Members were also working in the high-risk settings of aged and home care, and in health roles as paramedics, nurses and in pathology. Given the impacts of the pandemic, we delayed data collection for our project until it would be less of an imposition. Although members adapted quickly, the new working arrangements were not without significant challenges, and it became clear that the public health measures had led to WHS challenges relating to occupational heat. The delay in commencing also presented a chance to incorporate questions about Covid-19 into the project. This study is one of only a handful to examine the impact of the pandemic on occupational heat outside of health settings.[6] Moreover, studies to date have chiefly examined the issue as one that is technical or scientific in nature, which can be solved through improved protocols or PPE modifications. Such approaches fail to consider how these issues are fundamentally shaped by exploitation and contestation within the labour process.

Semi-structured interviews with 16 key informants took place from September 2020 to March 2021, and a survey (n=798) of 35 questions was open to members in the Allied Division of UWU in February 2021 (distributed by email in the final month of the Australian summer). A mixed methodology was chosen to gather rich data on how heat is produced within and through the work context, and to explore how workers attempt to manage this. The focus of the survey was high heat and climate change with two questions about the impact on Covid-19.[7] 773 participants answered two questions about whether

Covid-19 had: 1) impacted their experience of heat (closed) and, 2) if so, how (open). Several interviewees raised the impact of the pandemic, although they were not asked directly about it.

Covid-19 related measures and occupational heat risk

Mapping risk: Evidence from other studies

Emerging concerns around the nexus of heat-related illness and coronavirus transmission precautions were summarised early in the pandemic, in a technical brief published by the Global Health Heat Information Network (GHHIN).[8] They outlined that workers required to wear PPE and work in inadequately cooled conditions for prolonged periods are at increased risk of health effects, as well as prone to reduced cognitive capacity, productivity and wellbeing.[9] In conditions of forced isolation, individuals without access to proper cooling controls such as air-conditioning are also at great risk. The GHHIN noted that populations that are particularly vulnerable to Covid-19 are often also vulnerable to heat stress. This includes those with underlying health conditions, older people, and people living in isolated conditions.

The use of PPE, mandated in many workplaces and public spaces throughout the pandemic, increases an individual's susceptibility to heat-related illnesses, by making it more difficult to lose body heat, adding additional weight and restricting mobility.[10] While wearing PPE – such as face masks, a visor or goggles, gloves, impermeable aprons and hazmat suits – heat evaporation and heat exchange between the body and the environment is restricted.[11] The body's regular heat-loss mechanisms, such as sweating, cannot operate as efficiently, increasing the risk of overheating and thermal strain.[12] Heat stress can have secondary negative impacts upon cognitive performance,[13] work capacity and productivity,[14] and injury and accident rates.[15] PPE can also impair a wearer's dexterity, visibility and auditory response, raising safety risks and reducing performance.[16] Other measures designed to reduce the transmissibility of the virus have put additional strain on workers. For instance, during early Covid-19 outbreaks in India, air-conditioning was banned in some emergency departments.[17] Outdoor areas were erected to accommodate patient loads and mitigate risk, further exposing patients and workers to high heat conditions, in various international locations.[18]

Studies documented the serious impacts of these measures on the health, safety and performance of workers, which mostly relate to healthcare workers in health settings. Bongers et al. compared health workers conducting the same task before and after the introduction of pandemic-specific PPE measures.[19] They found heat strain was a major challenge, with female workers and those over 40 years old most effected, "whereas exposure time and sports activity level [as a proxy for fitness] were not significantly associated with heat strain prevalence."[20] In a study of nurses and bus drivers working in mobile Covid-19 screening units in Taiwan, Chang et al. found evidence that workers wearing PPE and working in hot weather outdoors had a significant decline in their kidney function.[21]

Davey and colleagues anonymously surveyed healthcare workers in the UK to evaluate their perceptions of heat stress during PPE use.[22] These workers were employed by the National Health Service UK and required to wear heavy PPE. The project found that almost all experienced increased sweating while wearing PPE, almost 90 percent felt "very uncomfortable" or "uncomfortable," and two thirds felt "hot." Headache was the most reported symptom of heat-related illness, followed by fatigue. PPE inhibited the physical performance of two thirds of workers, and over 90 percent said PPE made their job more difficult. The study authors concluded that modifications to PPE design or policy were urgently needed to reduce levels of heat stress.[23]

Jegodka et al. surveyed nurses and nursing assistants working in Germany, who wore a significant amount of Covid-19-related PPE.[24] On hot days, these workers reported that they sweated more (99.5 percent), had more difficulty breathing (93 percent), and had difficulty concentrating (85.8 percent). Heat-related symptoms were reported at significant rates, including exhaustion, irritability, headaches, dermatological problems and dizziness. Almost 90 percent of nurses reported that these conditions required more patient time. Despite these problems, protective workplace measures to deal with these challenges were limited – with less than ten percent benefiting from revisions in care planning, moving strenuous work to the cool morning or evening, more frequent or longer breaks, a shortened working day. Only 3.1 percent saw the employment of additional nursing staff. Heat protection plans were implemented in less than 1 percent of hospitals, 22.2 percent of nursing homes and 6.7 percent of ambulatory care services. Alarmingly, only 22.2 percent of ambulatory care service workers, 12.5 percent of nursing home workers and zero percent of hospital workers felt protected at work.

Lee et al. surveyed 165 healthcare workers in India and Singapore on workplace heat exposure adaptation and behavioural changes while working in PPE.[25] Many of these workers reported heat stress symptoms at work and that they had such high workloads that 65 percent were unable to take breaks, and 71 percent reported that hydration breaks were an inconvenience. Some healthcare workers were required to doff their PPE during breaks or meals and don a new set afterwards, and a majority reported that they avoided taking breaks to conserve PPE and avoided taking breaks to preserve infection control. Drinking and eating was avoided by more than a third of surveyed workers to reduce the number of toilet breaks they had to make.

There have been calls for further research to understand the potentiating impact of PPE usage, environmental factors, and workload, to improve occupational heat stress outcomes among working populations.[26]

Managing risk: Suggestions from other studies

Similar to studies of the effects of heat stress on working populations, recent studies (including some of the aforementioned studies) on the relationship between heat stress and Covid-19 management practices make suggestions for risk management and heat adaptation. National heat-health action plans are suggested by some researchers as an adaptation strategy, for example, HEAT-SHIELD in Europe or WORKLIMATE in Italy.[27]

Jegodka et al. conclude that "there is a distinct need for stronger institutional policies that acknowledge the existence of occupational heat stress and the need of cooling for nurses and nursing assistants," particularly in the context of higher annual average temperatures and more frequent heat waves driven by climate change.[28] They argue that "healthcare organizations should provide opportunities for nurses and nursing assistants to alleviate the thermal stress they experience, including institutional policies outlining a maximum duration for which PPE could be worn, at least during peak temperatures."[29] The authors stress that any heat protection measures introduced to protect nursing staff must be "stringently enforced."[30] Herstein et al. argue that the "appropriate work-rest cycle must be considered to prevent heat stress" and "generally, this entails recommendations for limited time in PPE followed by adequate rest outside of PPE but also must factor in what happens within patient care shifts."[31]

Lee et al. noted that while "work-rest balance may improve heat tolerance," "strategies aimed at ensuring work-rest balance or PPE removal might be unfeasible during a pandemic."[32] Pre-cooling strategies (such as through ice slurry ingestion) "may contribute to extended work tolerance times for HCWs wearing PPE during a pandemic."[33] As there is some evidence that improving aerobic fitness can help to guard against heat stress, the authors note that "it may behove employers of HCWs to take ownership of employee well-being and aerobic fitness to optimize HCWs' heat tolerance."[34] Davey et al. conclude that

> modifications to current PPE designs or policies on the use of PPE are required urgently [in the UK] to reduce the level of heat stress and the prevalence of other issues that jeopardise the performance, safety and well-being of HCWs with potential consequences for patients.[35]

The authors suggest that this may include the provision of cooler working environments and/or a reduction in the amount of time PPE is mandated for.

The research on the implications of high heat in the context of Covid-19 detailed here, and published more broadly, has three limitations. First, research has focused on the health sector, and we require more information about other settings. Second, investigations have most deeply explored PPE and we require greater exploration of other changes mandated by pandemic WHS protocols. Third, there has been a tendency to treat the problems workers have encountered as technical issues that can be solved simply through a change in protocols, the redesign of PPE or the strict enforcement of new protocols, and as such there is a failure to consider how questions of exploitation and control in labour process are imbricated. The following section details the experiences of the workers involved in this study, before discussing the implications from the data.

Heat, Covid-19 and UWU members

More than a third of those surveyed (37.5 percent) reported they had to make changes at work during the pandemic that affected their experience of the heat, while 62.5 percent said there had been no impact. For those that answered there was an impact, their open-ended question responses as to how were analysed and coded for the "key impact" and are detailed in Table 9.1. The negative impacts ranged from milder issues such as heat discomfort from wearing masks, through to significant problems stemming from intensified workloads or wearing full PPE in spaces with no air-conditioning. Some changes led to new heat stress issues that had not existed previously; for example, social distancing and changes to work tasks led to increased time outside (such as on lunch breaks) or needing to perform extra work like school cleaning in the hottest part of the day (when this is usually reserved for cooler periods).

The most common problems with increased heat stress were associated with PPE, including the use of gowns, aprons, gloves, goggles, face masks, shields

TABLE 9.1 Main impact of COVID-19 on work duties indicated in response

Number of respondents 254	*Number raised issues*	*Percentage total respondents*
PPE	201	79.1%
Time outside	21	8.3%
Extra duties (compliance)	23	9.1%
Extra duties (general)	8	3.1%
Restricted space (general)	7	2.8%
Restricted space (loss of air con)	7	2.8%
Ban on using air con/showers	6	2.4%
Intensity of work	6	2.4%
Changes to roster	5	2.0%
WFH	4	1.6%
Extra cost	4	1.6%
Insufficient staff	4	1.6%

and balaclavas. Members using PPE advised they experienced heat, fatigue, sweating, skin rashes and difficulty breathing. Members described wearing PPE during high heat for an entire shift as "debilitating" and "stifling," noting that wearing PPE "adds to the effects of the heat."

The experience of PPE was at times compounded by altered or extended shift times, and higher workloads, especially for essential workers such as cleaners and healthcare personnel. As one member working in pathology collection in Queensland, a large tropical and subtropical state with hot and humid summers, told us:

> PPE is so hot it makes you feel just horrible and there is no relief from it for the whole shift. Bottled water is given to workers, but breaks are hard to get because staffing is low, and demand is overwhelmingly high.

Members also reported that changes to shift times to stagger the numbers of workers on site resulted in outdoor work being completed during the middle of the day rather than in the morning or evening, and an increase in higher intensity work such as the cleaning of outdoor play equipment and drinking fountains in educational settings. Some members advised they were not provided with additional time to complete higher workloads, or the ability to take extra breaks to account for the heat effects of working in PPE. Reduced access to air-conditioned spaces was also raised. Restrictions on the number of people allowed in break rooms due to social distancing regulations were a problem, as this forced workers to take their breaks outside or seek other spaces without cooling to rest in. This included a small number of people directed to work from home who did not have effective cooling or could not afford the expense of running air-conditioners all day.

A pathology courier in Queensland noted that the "extra time required to adhere to covid protocols [was] not factored into daily routines." Another health worker who worked in a mobile vehicle stated "we are now wearing long sleeved gowns, masks, and shields. The air cons in the vans are at least 40 years old, and really don't do the job." A Northern Territory based paramedic made the point that cooling and ventilation can be a problem in the industry generally, and that PPE can compound problems of heat stress: "Wearing of PPE in patient's poorly ventilated or non-air-conditioned houses. Decontamination of vehicles whilst wearing full PPE including gown, goggles [and] gloves on top of heavy wearing uniform. Wearing this PPE for long periods of time." While some members reported increased access to water and other items to assist them, others said there was no access to water, that it was difficult to consume it or take a break, and that access to cooling showers had been stopped. Some workers avoided drinking water because it meant they had to get in and out of PPE, as it involved extra physical labour in hot conditions.

Cleaners in a range of education, health and other settings described significantly increased workloads. Several noted that some tasks had to be done outside in the hottest parts of the day, such as cleaning play equipment, which were tasks usually left for cooler times of the day to maximise safety and comfort. One school cleaner described how heavy indoor work, like vacuuming, had to be completed without air-conditioning for a period in the pandemic, as the safety of leaving ventilation running was unclear. Cleaners in meatworks found the use of heavy PPE, including full coverage and plastic balaclavas in what is a high-risk transmission situation, particularly difficult to cope with. One of these workers based in the state of Western Australia noted that:

> wearing [a] plastic apron, face mask, eye protection and gloves makes me sweat profusely, causing dehydration, fatigue and headaches. It is impossible to work without a higher risk of causing an injury and [I] cannot perform at an efficient level of competence. Goggles make it hard to see due to fogging up.

This worker was not alone in pointing to other WHS problems created by new forms of PPE aside from the direct increase in body temperature, such as sweating under gloves making it difficult to handle equipment and perform tasks. Another worker noted masks caused fogging and this was dangerous while driving machines while wearing glasses for vision. Another noted that PPE masks limited peripheral vision, and that the "lower scope of vision making moving up and down steps hazardous." A veterinary nurse pointed out various complications such as:

> Having to consult with the public outdoors instead of indoors, having to bring animals in without owners means more walking, heavy lifting, more time pressures so having to run instead of walk. Ongoing understaffing issues exacerbate all of the above.

Several workers indicated PPE had increased heat stress and caused cognitive problems for them. A laundry worker in Western Australia, which is the country's largest state in terms of land mass with a climate ranging from Mediterranean in the south to tropical in the north, reported that

> Wearing a mask at work, and with the heat and plastic shields put in place to protect us from the 1.5 metre rule, I've experienced fatigue, brain fogginess, and become extremely flustered with the heat, due to little circulation in some areas.

Much of Western Australia, even without PPE, has a very difficult climate to tolerate during summer. Finally, in one upsetting outcome, a health sector worker reported that "a patient died from heat stress while having to wait … in their car, due to Covid restrictions." The psychosocial impacts of the experiences that workers reported to us were very significant.

Workers noted the costs of cooling they faced, and that employers more readily funded items needed related to the pandemic but not in relation to occupational heat. A home care worker[36] based in New South Wales told us how they and some clients could not afford to run their air-conditioners, and how employers failed to manage occupational heat:

> Yep. I don't run my air-conditioner, because my car cost me too much to run already … look, we don't get paid all that much, so I know it sounds really tight. But um, you got to be careful of what you spend … Some client's places – well, they won't put on air-conditioner. So, it can be very hot. Especially when we give showers we've got steam in the showers that they won't allow us to open up windows – nearly pass out from the steam that's happening, which also robs the oxygen in the room with so much steam, and especially with wearing masks with Covid and also goggles or a face shield, and also your gloves and also your plastic apron over our plastic clothes, which are polyester, which is a huge issue.
>
> I think there should be much more onus on the employer, to you know … be providing … cold packs to staff you know, providing, like even in Covid they did it, you know, providing us with thermometers and stuff providing us with tools so we can do our jobs, or fans and things. So, providing us with you know, more tools to help us survive.

"Funny stories aside," began another home care worker caring for disabled people based in Western Australia, as she related a story of when she raised the possibility of needing to plan for the impacts of major events related to climate change:

> at the end of [2020], we had our final meeting for the year … and the team leader, Sarah [said], 'Has anybody else got any questions? Any other issues?' And I don't know why, it just popped into my head, so I said, 'Well, what happens if for some reason we can't get out to visit our clients, you know, what if something happens [and] they're stuck at home and we can't get out?' Now I was thinking of a fuel crisis or a bushfire, but actually we had the pandemic. So, it actually came true, and there was no plan, because they all laughed, and said, 'no, that would never happen.' But climate change is something that I think they should be planning for. If it gets to the point where workers can't safely go out during the summer, how are we going to look after people in their homes if we can't even leave our own homes

> because it's too hot? I don't think anybody's actually really thought that far ahead. Because they will they're like, 'oh, that won't happen,' but I could see it happening in a few years the way things are going.

A finding from the broader UWU project was that procedures related to managing high heat are on the whole poor.[37] This included the failure of employers to establish procedures, as well as implementing them where they exist. The research also revealed that the workplaces studied were generally unprepared for novel WHS developments. This was also the case during the Australian Black Summer bushfires of 2019 and 2020, when thick smoke and ash enveloped four-fifths of Australia's population, killing 33 people directly and 455 people indirectly as a result of air pollution.[38] Both the pandemic and the fires evidence the fact many employers do not prioritise planning for extreme WHS situations in advance or have clear processes for dealing with urgent new circumstances.

Conclusion: Contestation in the labour process

The UWU project confirmed the well-established relationship between temperature, humidity and work effort in producing heat stress. It demonstrated that while workers experienced heat stress outside of mandated Covid-19 protocols, pandemic procedures both exacerbated occupational heat related risks as well as created new challenges and that these were not well managed. The research revealed inadequacies in current WHS management frameworks, demonstrated workplaces were largely unprepared for the pandemic emergency, and that employers failed to ensure workers' wellbeing in relation to occupational heat. In this study the tension between pandemic measures and WHS protocols in relation to occupational heat largely resulted in the former trumping the latter.

While occupational heat was a problem for the majority of workers in this study prior to the pandemic, established ways to manage these risks (collective and individual) were undermined by the implementation of Covid-19 protocols and the failure of employers to adapt to growing heat stress. For example, school cleaners in Queensland had been allowed to complete heavier labour outside of the heat of the day, and complete work indoors in air-conditioning. Mandatory protocols changed this and required the cleaning of play equipment and drinking fountains more often and during peak temperatures; yet workers reported they were not given extra time to complete tasks or take more rest, and nor were extra workers hired. This situation is unlike the health settings examined overseas, where virus spread reduced staff numbers and impacted the ability to deliver services. While some sectors experienced labour shortages because overseas workers were unable to enter the country, high unemployment due to the pandemic resulted in a large reserve army of workers keen for many jobs undertaken by the workers involved in this study.

It was not that these employers could not act to address increased occupational heat issues, but that they didn't. In comparing workers from a single union in both health and other settings, the data demonstrates that employers failed to act to eliminate or mitigate the increased occupational heat impacts related to the pandemic in a wide variety of settings. Employers appeared to be significantly driven by productivity and cost concerns, even when health risks were clear and mitigating action was possible. This finding is also supported by what UWU members advised occurred in the pre-pandemic 2019–2020 Black Summer bushfires. Several UWU members noted that employers failed to act to eradicate or mitigate the WHS impacts related to smoke and ash pollution, and that members were forced to continue working despite risks. One member, a quality assurance officer and process worker working at a manufacturing site in Western Sydney, stated they just "worked through it." They were instructed to wear an unsuitable mask to prevent them breathing in "sticky smoke," and he described the experience as "kinda dystopic."

To date, problems of increasing heat risk related to Covid-19 protocols have largely been framed as resulting from pressures that could not easily be planned for or were peculiar to the pandemic. Authors argue that in the future similar issues can be addressed through better planning, or might simply have been unavoidable because of the particular pressures generated by the pandemic itself (such as increasing workloads in health settings and labour shortages related to workers needing to isolate). Yet, mandatory PPE or uniform requirements, shift patterns and lengths, workers' access to breaks, the availability of cooling mechanisms, and so on, are contested industrial relations matters connected to control over the labour process. In the case of some UWU workers, there are further difficulties in occupations where work is "distributed" in nature, such as for pathology collectors, delivery people, cleaners and care workers working from their cars and in people's homes. This brings additional complexities and challenges that are not often discussed in relation to work and climate, which can tend to focus on workers within stand-alone workplaces. In addition, processes are not always governed in detail by legislation and regulation, and even where they are, they are ultimately enforced (or not) by employers and complied with or contested by labour.

The interests of labour and employers in WHS are not alike. Despite rhetoric that everyone is responsible for safety, the parties have unequal power and control in the labour process and workers cannot always act to ensure their own safety. For this reason, and as others have argued, WHS management should be understood as a social – and not just technical – phenomena, embedded in relations of production.[39] Given "exceptional" WHS circumstances will become the norm as the consequences of climate change deepen, labour organisations will need to be particularly attentive to how moments of crisis can undermine past gains. Alongside this there is a desperate need to (re)build worker agency to ensure labour can take action to protect themselves, and force employers to act in relation to the deepening impacts of global heating.

Ultimately, climate change adaptation in the workplace must be understood as a thoroughly political process, where vulnerability and exploitation are reproduced and contested.

Notes

1 This work was supported by a grant from the United Workers Union. I would like to acknowledge the work and collegiality of Freya Newman who provided research assistance on this project.
2 Ebi et al., "Hot Weather and Heat Extremes: Health Risks."
3 Hyatt, Lemke, and Kjellstrom, "Regional Maps of Occupational Heat Exposure: Past, Present, and Potential Future."
4 Quinlan, "Five Challenges to Humanity: Learning from Pattern/Repeat Failures in Past Disasters?"
5 Australia has six states, and two territories run by the Federal Government. During Covid-19 outbreaks, controversial, state-led measures to control virus spread included: the order put in place on public housing residents in an inner Melbourne high rise, and restrictions on residents in several local government areas in Sydney. Because the virus spread (limited as it was) occurred amongst workers doing essential and low paid roles such as security, clearing and food logistics, those most impacted by these restrictions were on lower-than-average incomes and living in locations with a high percentage of racial diversity. This meant the restrictions effectively targeted poor and non-white populations.
6 Bonafede et al., "Workers' Perception [of] Heat Stress: Results from a Pilot Study Conducted in Italy during the COVID-19 Pandemic in 2020"; Santurtún and Shaman, "Work Accidents, Climate Change and COVID-19."
7 The survey had a 4.5 percent response rate. Questions were not compulsory, giving members the ability to opt out of any question they did not want to respond to, or that was not relevant to them, and still complete.
8 GHHIN, "Protecting Health from Hot Weather during the Covid-19 Pandemic."
9 GHHIN, "Protecting Health from Hot Weather during the Covid-19 Pandemic," 4.
10 Davey et al., "Heat Stress and PPE during COVID-19: Impact on Healthcare Workers' Performance, Safety and Well-Being in NHS Settings."
11 Kuklane et al., "Ebola."
12 Daanen et al., "COVID-19 and Thermoregulation-Related Problems: Practical Recommendations"; Lee et al., "Heat Stress and Thermal Perception amongst Healthcare Workers during the COVID-19 Pandemic in India and Singapore"; Potter, Gonzalez, and Xu, "Ebola Response."
13 Zhang, de Dear, and Hancock, "Effects of Moderate Thermal Environments on Cognitive Performance."
14 Chinnadurai et al., "Influence of Occupational Heat Stress on Labour Productivity"; Kjellstrom et al., "The Direct Impact of Climate Change on Regional Labor Productivity"; Morabito et al., "Heat-Related Productivity Loss."
15 Xiang et al., "Association between High Temperature and Work-Related Injuries."
16 Davey et al., "Heat Stress and PPE during COVID-19"; Foster et al., "Occupational Heat Stress and Practical Cooling Solutions for Healthcare and Industry Workers"; Herstein et al., "A Pilot Study of Core Body Temperatures in Healthcare Workers."
17 Lee et al., "Heat Stress and Thermal Perception amongst Healthcare Workers."
18 Lee et al., "Heat Stress and Thermal Perception amongst Healthcare Workers."
19 "Heat Strain and Use of Heat Mitigation Strategies among COVID-19 Healthcare Workers."
20 Bongers et al., 7.

21 "Mobile COVID-19 Screening Units."
22 Davey et al., "Heat Stress and PPE during COVID-19."
23 Davey et al., "Heat Stress and PPE during COVID-19," 188.
24 Jegodka et al., "Hot Days and Covid-19."
25 Lee et al., "Heat Stress and Thermal Perception amongst Healthcare Workers."
26 See Herstein et al., "A Pilot Study of Core Body Temperatures in Healthcare Workers."
27 See https://www.heat-shield.eu/ and https://www.worklimate.it/.
28 Jegodka et al., "Hot Days and Covid-19," 5.
29 Jegodka et al., "Hot Days and Covid-19."
30 Jegodka et al., "Hot Days and Covid-19."
31 "A Pilot Study of Core Body Temperatures in Healthcare Workers Wearing Personal Protective Equipment in a High-Level Isolation Unit," 433.
32 Lee et al., "Heat Stress and Thermal Perception amongst Healthcare Workers," 9.
33 Lee et al., "Heat Stress and Thermal Perception amongst Healthcare Workers," 10.
34 Lee et al., "Heat Stress and Thermal Perception amongst Healthcare Workers."
35 Davey et al., "Heat Stress and PPE during COVID-19," 188.
36 Home care workers are particularly poorly paid in Australia and often work on precarious employment contracts, mostly working with disabled and elderly clients living on meagre government benefits.
37 Humphrys, Goodman, and Newman, "'Zonked the Hell Out'"; Humphrys and Newman, "High Heat and Climate Change at Work."
38 Wahlquist, "Australia's Summer Bushfire Smoke Killed," 445.
39 Creighton and Gunningham, "Is There an Industrial Relations of Occupational Health and Safety?"; Elling, "The Political Economy of Workers' Health and Safety"; Heino, "The State, Class and Occupational Health and Safety."

Bibliography

Bonafede, Michela, Miriam Levi, Emma Pietrafesa, Alessandra Binazzi, Alessandro Marinaccio, Marco Morabito, Iole Pinto, et al. "Workers' Perception [of] Heat Stress: Results from a Pilot Study Conducted in Italy during the COVID-19 Pandemic in 2020." *International Journal of Environmental Research and Public Health* 19, no. 13 (2022): 8196.

Bongers, Coen C. W. G., Johannus Q. de Korte, Mike Zwartkruis, Koen Levels, Boris R. M. Kingma and Thijs M. H. Eijsvogels. "Heat Strain and Use of Heat Mitigation Strategies among COVID-19 Healthcare Workers Wearing Personal Protective Equipment – A Retrospective Study." *International Journal of Environmental Research and Public Health* 19, no. 3 (2022): 1905.

Chang, Teng-Hsiang, Ching-Yu Lin, Jason Kai Wei Lee, Che-Jui Chang, Wan-Chin Chen and Hsiao-Yu Yang. "Mobile COVID-19 Screening Units: Heat Stress and Kidney Function Among Health Care Workers." *American Journal of Kidney Diseases* 80, no. 3 (2022): 426–28.

Chinnadurai, Jeremiah, Vidhya Venugopal, P. Kumaravel and R. Paramesh. "Influence of Occupational Heat Stress on Labour Productivity – A Case Study from Chennai, India." *International Journal of Productivity and Performance Management* 65, no. 2 (2016): 245–55.

Creighton, Breen and Neil Gunningham. "Is There an Industrial Relations of Occupational Health and Safety?" In *The Industrial Relations of Occupational Health and Safety*, edited by Breen Creighton and Neil Gunningham, 3–10. Sydney: Croom Helm Australia, 1985.

Daanen, Hein, Stephan Bose-O'Reilly, Matt Brearley, D. Andreas Flouris, Nicola M. Gerrett and Huynen Hunter M. Jones, et al. "COVID-19 and Thermoregulation-Related Problems: Practical Recommendations." *Temperature* 8, no. 1 (2021): 1–11.

Davey, S. L., B. J. Lee, T. Robbins, H. Randeva and C. D. Thake. "Heat Stress and PPE during COVID-19: Impact on Healthcare Workers' Performance, Safety and Well-Being in NHS Settings." *Journal of Hospital Infection* 108 (2021): 185–88.

Ebi, Kristie L, Anthony Capon, Peter Berry, Carolyn Broderick, Richard de Dear, George Havenith, Yasushi Honda, et al. "Hot Weather and Heat Extremes: Health Risks." *The Lancet* 398, no. 10301 (2021): 698–708.

Elling, Ray H. "The Political Economy of Workers' Health and Safety." *Social Science and Medicine* 28, no. 11 (1989): 1171–82.

Foster, Josh, Simon G Hodder, James Goodwin and George Havenith. "Occupational Heat Stress and Practical Cooling Solutions for Healthcare and Industry Workers During the COVID-19 Pandemic." *Annals of Work Exposures and Health* 64, no. 9 (2020): 915–22.

GHHIN. "Protecting Health from Hot Weather during the Covid-19 Pandemic." Technical Brief. Global Heat Health Information Network (GHHIN), 27 May 2000. https://ghhin.org/wp-content/uploads/technical-brief-COVID-and-Heat-finalv2-1.pdf

Heat-shield. 2016. https://www.heat-shield.eu/

Heino, Brett. "The State, Class and Occupational Health and Safety: Locating the Capitalist State's Role in the Regulation of OHS in NSW." *Labour and Industry* 23, no. 2 (2013): 150–67.

Herstein, Jocelyn J., Abdoulaziz A. Abdoulaye, Katelyn C. Jelden, Aurora B. Le, Elizabeth L. Beam, Shawn G. Gibbs, Angela L. Hewlett, et al. "A Pilot Study of Core Body Temperatures in Healthcare Workers Wearing Personal Protective Equipment in a High-Level Isolation Unit." *Journal of Occupational and Environmental Hygiene* 18, no. 9 (2021): 430–35.

Humphrys, Elizabeth, James Goodman and Freya Newman. "'Zonked the Hell Out': Climate Change and Heat Stress at Work." *Economic and Labour Relations Review* 33, no. 2 (2022): 256–71.

Humphrys, Elizabeth and Freya Newman. "High Heat and Climate Change at Work: Report for the United Workers Union." Ultimo: Climate Justice Research Centre, University of Technology Sydney, September 2021.

Hyatt, Olivia M., Bruno Lemke and Tord Kjellstrom. "Regional Maps of Occupational Heat Exposure: Past, Present, and Potential Future." *Global Health Action* 3, no. 1 (2010): 5715.

Jegodka, Yvette, Lena Lagally, Hanna Mertes, Katharina Deering, Julia Schoierer, Barbara Buchberger and Stephan Bose-O'Reilly. "Hot Days and Covid-19: Online Survey of Nurses and Nursing Assistants to Assess Occupational Heat Stress in Germany during Summer 2020." *The Journal of Climate Change and Health* 3 (2021): 100031.

Kjellstrom, Tord, R. Sari Kovats, Simon J. Lloyd, Tom Holt and Richard S. J. Tol. "The Direct Impact of Climate Change on Regional Labor Productivity." *Archives of Environmental and Occupational Health* 64, no. 4 (2009): 217–27.

Kuklane, Kalev, Karin Lundgren, Chuansi Gao, Jakob Löndahl, Elisabeth Dalholm Hornyanszky, Per-Olof Östergren, Per Becker, et al. "Ebola: Improving the Design of Protective Clothing for Emergency Workers Allows Them to Better Cope with Heat Stress and Help to Contain the Epidemic." *The Annals of Occupational Hygiene* 59, no. 2 (2015): 258–61.

Lee, Jimmy, P. K. Vidhya Venugopal, Sharifah B. Latha, Clarence H. Alhadad, Nicholas Y. Leow, Esther Tan Goh, Tord Kjellstrom, Marco Morabito and Jason K. Lee. "Heat Stress and Thermal Perception Amongst Healthcare Workers During the COVID-19 Pandemic in India and Singapore." *International Journal of Environmental Research and Public Health* 17, no. 21 (2020): 8100.

Morabito, Marco, Alessandro Messeri, Alfonso Crisci, Junzhe Bao, Rui Ma, Simone Orlandini, Cunrui Huang and Tord Kjellstrom. "Heat-Related Productivity Loss: Benefits Derived by Working in the Shade or Work-Time Shifting." *International Journal of Productivity and Performance Management* 70, no. 3 (2021): 507–25.

Potter, Adam W., Julio A. Gonzalez and Xiaojiang Xu. "Ebola Response: Modeling the Risk of Heat Stress from Personal Protective Clothing." *PLOS ONE* 10, no. 11 (2015): e0143461.

Quinlan, Michael. "Five Challenges to Humanity: Learning from Pattern/Repeat Failures in Past Disasters?" *Economic and Labour Relations Review* 31, no. 3 (2020): 444–66.

Santurtún, Ana and Jeffrey Shaman. "Work Accidents, Climate Change and COVID-19." *Science of the Total Environment* 871 (2023): 162129.

Wahlquist, Calla. "Australia's Summer Bushfire Smoke Killed 445 and Put Thousands in Hospital, Inquiry Hears." *The Guardian*, 26 May 2020. https://www.theguardian.com/australia-news/2020/may/26/australias-summer-bushfire-smoke-killed-445-and-put-thousands-in-hospital-inquiry-hears

Worklimate 2.0. 2024. https://www.worklimate.it/

Xiang, Jianjun, Peng Bi, Dino Pisaniello, Alana Hansen and Thomas Sullivan. "Association Between High Temperature and Work-Related Injuries in Adelaide, South Australia, 2001–2010." *Occupational and Environmental Medicine* 71, no. 4 (2014): 246.

Zhang, Fan, Richard de Dear and Peter Hancock. "Effects of Moderate Thermal Environments on Cognitive Performance: A Multidisciplinary Review." *Applied Energy* 236 (2019): 760–77.

10

INDUSTRIAL CONTROL AND CAPITALIST INEFFICIENCY

Manufacturing Workforce Development in the Polycrisis

Mark Dean and Daniel Nicholson

Introduction

Since its election in 2022, two of the central concerns of the Australian Labor government – under Prime Minister Anthony Albanese – have been developing a more ambitious policy agenda for Australia's response to climate change, and bolstering the country's military capabilities in response to real and perceived risks of regional conflict. The risks that emerge from the deepening ecological crisis and increased geopolitical tensions in the region interact in ways that mean the overall impacts exceed the sum of each part at various and often unpredictable junctures – a facet of what historian Adam Tooze has referred to as the "polycrisis."[1] In response to these overlapping policy challenges, the Federal Government has committed to a renewed industry policy, pledging to invest Australia's sovereign capability to manufacture defence materiel and green energy infrastructure. However, several decades of neoliberal economic policy, characterised in part by an absence of any coordinated industrial policy, have left Australia's domestic manufacturing capabilities and skilled workforce severely depleted.[2] This chapter is concerned with the workforce development barriers the government faces to tackling the polycrisis by deploying *Future Made in Australia* (FMiA), its key industrial policy response.

Announced in early 2024, the *Future Made in Australia Act* promises to provide an overarching industry policy framework to bring together several of the government's existing initiatives and special investment vehicles.[3] Reflecting the two central motivations for Australia's renewed interest in industrial development, the government announced two "priority streams" for investment. First, the "Net Zero Transformation Stream," which will emphasise areas of domestic comparative advantage in the transition to a net zero economy such

DOI: 10.4324/9781003512240-13

as green metals. Second, the "Economic Resilience and Security Stream" will include industries where sovereign capability is considered essential to national security and economic resilience.[4] The lofty aims of FMiA will require Australian industry to integrate a range of new technologies and to develop significant new capabilities. In the early stages of consultation around this new, overarching policy, trade unions have reiterated concerns raised about the ability of Australian industry initiatives to deliver on the government's stated aims without a coordinated approach to workforce and skill development.

This chapter argues that a managerial tension between developing industrial capabilities through worker upskilling and empowerment on the one hand, and ensuring managerial control over the production process through worker deskilling on the other, represent a significant roadblock to realising the goals of Australian industry policy. To make this argument we draw on the experiences of the Australian Manufacturing Workers' Union (AMWU) and Australian Shipbuilding Federation of Unions (ASFU) around workforce development in the naval shipbuilding programme. We demonstrate that the managerial tension between development and control has restricted the capacity of industry to integrate advanced manufacturing technologies and scale productive output and limited the potential positive "spillover" effects of industry policies for other sectors of manufacturing. We focus our empirical analysis on the defence sector because industry policy is more mature here. However, we argue that the shortcomings of government industry policy in defence is likely to be replicated in the pursuit of a "green manufacturing" sector, unless the tension between managerial control and industrial development can be ameliorated.

The central contribution of this chapter is to highlight that for industry policy to play a major, successful role in facilitating Australia's response to the "polycrisis," intervention in the normal operation of capitalist markets at multiple levels is required. As evidenced by the AMWU and ASFU's experience in naval shipbuilding, government investment will not be enough. The drive of capitalist enterprises in the defence sector to maximise profits by maintaining tight control over production processes has stymied both productivity gains from technological change and the development of the skilled manufacturing workforce Australia needs for the renewable energy transition. The outcome will be replicated in green manufacturing unless a combination of trade union and government interventions can force industry to invest in skills and training and allow workers a greater degree of control over their work processes. Ultimately, we believe that lessons from defence industry policy suggest this tension represents a significant threat to the government's ambitions to drive Australia's transition away from fossil fuels using industry policy. Australia's ability to effectively deploy industry policy to tackle overlapping crises of climate change, increasing geopolitical tensions and ever more disrupted global supply chains will depend on the development of a skilled manufacturing workforce capable of doing that work – the managerial drive for workplace control has the potential to prevent the realisation of these goals.

The chapter begins with a brief discussion of the geo-political and environmental contexts that have brought such a central focus onto Australia's manufacturing capabilities. It then situates this chapter in the literature around industry development and labour process analysis, demonstrating the managerial tension between upskilling workers for greater productive capacity and industrial development, and deskilling workers to maximise managerial control. Next, we discuss trade union advocacy around workforce development through industry policy, vocational education and training, and the empowerment of workers on the shopfloor, and demonstrate managerial resistance to this agenda. We then argue that while employers generally share union concerns with Australia's broader sectoral and industrial development, their desire to maintain high levels of control over their workforce on the shopfloor causes them to resist worker empowerment and upskilling at multiple levels. We argue that this dynamic represents a manifestation of capitalist inefficiency, and a significant hurdle to Australia developing the capacity to build complex technologies like naval submarines, and renewable energy infrastructure like green steel and towers for offshore wind.

Australia's industrial capabilities and the polycrisis

The term "polycrisis" first came into use in academic circles in the 1970s, but its contemporary use has been mainstreamed by historian Adam Tooze who uses the term to describe the coming together of multiple calamities, and the ways in which these crises overlap and worsen one another.[5] In Australia, two central manifestations of the global polycrisis are increasing regional geopolitical tensions and the acceleration of the ecological crisis. These crises interact in complex ways. For example, the accelerating impacts of the environmental crisis have the potential to worsen Australia's security situation. As political scientist Tobias Ide put it, climate change has the potential to:

> undermine Australia's national security by disrupting critical infrastructure, by challenging the capacity of the defence force, by increasing the risk of domestic political instability in Australia's immediate region, by reducing the capabilities of partner countries in the Asia-Pacific region, and by interrupting important supply chains.[6]

This section briefly discusses how these two overlapping issues have spurred a renewed governmental interest in domestic manufacturing capabilities, and some of the hurdles faced to reindustrialising in the face of the polycrisis.

For around thirty years, Australia's Asia-focused economic strategy, tying its economic fortunes to the "Chinese Miracle," has helped Australia sustain the longest period of uninterrupted economic growth of any developed country in history.[7] It is not without irony then, that Australia's diplomatic, defence and security interests are deeply aligned with those of the USA, which views China as a real or potential antagonist. Since the early 2010s the "pivot to Asia" strategy of the Obama Administration – an attempt to curtail China's rising diplomatic and military influence in the Asia-Pacific – Australian governments have complied with the USA's foreign policy priorities and renewed military interests in the Asia-Pacific. Historian Clinton Fernandes describes Australia's foreign policy as placing it in a subservient position to its larger ally.[8] This leaves Australia in a situation whereby the aim of defence strategy "is to uphold US military dominance, hence the deliberate choice not to develop a separate Australian military strategy."[9] Rounding out this argument, Fernandes explains that a nation cannot develop a strategic industrial policy without foreign policy independence. On the contrary, Australia will be required to develop industrial capabilities in a way that fulfils its sub-imperial role to sustain US hegemony in the Asia-Pacific region.

With its economic fortunes tied to China, but its foreign policy priorities subordinated to US priorities, the rise of China and relative decline of the US has resulted in major defence industry policy challenges for Australia. These challenges were made explicit in the federal government's *Defence Strategic Review* (DSR), released in 2023, which outlined six key priorities for Australia's defence forces:

- the acquisition of nuclear-powered (but not nuclear-armed) submarines;
- the manufacture of precision munitions for striking longer-range targets;
- investments in defence infrastructure to enable force projection from Australia's northern bases;
- improving the growth and retention of a highly skilled defence workforce;
- translating disruptive technologies into capabilities in close partnership with Australian industry; and
- deepening diplomatic and defence partnerships with key partners in the Indo-Pacific.[10]

Without any mention of China, these priorities are clearly geared towards projecting Australian power North. Developing greater sovereign defence industrial capabilities thus presents itself as a central plank of Australia's geopolitical objectives. But the Australian defence industry has long faced obstacles to living up to various governments' geopolitical goals. Defence procurement, for example, has a history of significant cost overruns.[11] The DSR industrial priorities – were they all to be manufactured here, as the government notionally hopes will be the case – would represent a major leap forward in technological and industrial complexity. The AUKUS nuclear submarines, for example, are an industrial undertaking unlike anything Australia has previously undertaken. An estimated AUD$368 billion is to be spent over the next several decades to develop a domestic industrial base that can build, maintain and repair nuclear-powered submarines. The Department of Defence has also flagged a potential 50 percent cost blowout, with the program designed to augment the industrial capabilities of the USA and UK in the near-term.[12]

Before steel can be cut on the first Australian-built nuclear submarine, the Australian government and the Department of Defence has already been grappling with significant workforce development issues. The DSR has noted that: "Defence is facing significant workforce challenges … a recurring theme across all areas [of the Review] … This is an acute issue for Defence and is reflective of broader national challenges."[13] Numerous submissions to the DSR highlighted the shortage of highly qualified and experienced manufacturing workers to deliver defence industrial priorities,[14] reflecting a shortage of skilled manufacturing workers in the broader economy. Despite this, the DSR offers little in the way of detail about how the defence industry workforce will be developed and sustained. And, as will be discussed below, defence contractors have been reticent to develop these capabilities themselves, despite being rhetorically committed to doing so.

Issues around workforce development and general industrial capabilities have increasingly called into question the government's commitment to building nuclear submarines (and other defence materiel) onshore. Indeed, it was recently announced that Australia will purchase between three and five of the promised AUKUS submarines from the US, rather than building them in South Australia, due to a lack of local workforce and industrial capabilities.[15] These submarines will be a mixture of new and second-hand boats made in the USA – significantly bolstering US defence manufacturing, but doing little to build Australian capabilities or create local jobs. This move has sparked significant speculation as to whether Australia will, in fact, build any of its future submarine fleet domestically, or if Australian money will simply be used to augment the capabilities of US and UK defence contractors.[16] In fact, former foreign minister Alexander Downer recently described the idea that Australia would build submarines in South Australia as "a fairytale," saying "it will never happen."[17]

The difficulties faced by Australia's defence industrial policy programmes bring into sharp relief the federal government's plans to deploy industrial policy as a central aspect of its plan to meet international obligations to reduce Australia's carbon emissions. Historically, Australia's economic prosperity has been heavily underpinned by fossil fuel extraction and carbon-intensive industries such as mining. As recently as 2019, Australia was the world's sixth-largest producer of fossil fuel energy and the fourth-largest fossil fuel exporter.[18] Importantly from a political perspective, Australia's legacy fossil fuel regions are often heavily concentrated geographically – especially coal mining communities.[19] These communities are faced with significant economic and social dislocation by Australia's energy transition unless new industries can be developed, and workers redeployed into new sectors of the economy.

In the context of this tension between communities dependent on fossil fuel extraction for their livelihoods and the lived experience of climate calamities, Australia has struggled to develop a coherent policy to address climate change. Divisive "climate wars" plagued federal politics for almost twenty years, claiming the careers of many senior politicians and damaging Australia's diplomatic standing.[20] For over a decade, the AMWU has worked to build community support for a just transition in legacy fossil fuel communities through industry policy. This advocacy has occurred through community coalitions like the Collie Just Transition Working Group in Western Australia and the Hunter Jobs Alliance in New South Wales. These campaigns aim to build community support for ambitious climate action through developing good local jobs in green manufacturing.[21] Abundant access to solar, wind, and wave energy sources have spurred the Commonwealth Scientific and Industrial Research Organisation (CSIRO) to tout Australia as a "renewable energy super-power in waiting."[22] AMWU campaigns have hoped to pressure governments into developing a policy approach that harnesses these resources to create tens of thousands of good manufacturing jobs.

Encouragingly, the announcement of the FMiA suggests the federal government now views industrial development as a major plank of its policy response to the ecological crisis. Central to the government's political vision is that the creation of manufacturing and green energy jobs – especially in Australia's legacy fossil fuel regions – will be enough to maintain public support (sometimes referred to as "social licence") for a more ambitious energy transition. However, without coordinated workforce development around government-backed green manufacturing and energy projects there is a risk that the government's new industry policy will struggle to create local jobs and industrial capabilities.

In the following section, we briefly discuss a conceptual framework for understanding the tensions faced by managers and firms between industrial development on the one hand, and control of the workforce on the other. This is then used to discuss the failures of Australia's defence industry policy to generate promised employment and industrial capability outcomes. Finally, this experience is applied to the emerging green-industrial agenda, where we argue that Australia's energy transition risks failing unless unions and governments can build a coordinated workforce development programme for the new economy.

Industrial development, control and capitalist production

Central to the Defence Strategic Review's (DSR) vision for Australian defence industry is the adoption of radically new technologies both in terms of military hardware – notably long-range missiles and nuclear submarines – and advanced manufacturing technologies to produce this hardware. Additionally, the DSR notes the need to develop a manufacturing workforce capable of integrating these technologies in terms of manufacturing and maintenance. In a similar vein, the "Net Zero Transformation Stream" foregrounds the adoption of a raft of new production technologies and processes to facilitate Australia's transition to a low carbon economy. For example, the FMiA flags the development of a green metals industry in Australia, noting that significant "green iron, steel, alumina and aluminium will be required to support the energy transition, globally and domestically."[23] Developing a local green metals industry will involve the adoption of a raft of new technologies and building out substantial new supply chains. In both cases, integrating new technologies will require training new workers upskilling existing manufacturing workers in a range of new tasks, skills and competencies.

Some authors – especially in the heady world of management consultancy reports – have framed contemporary technology change in the manufacturing industry in a highly deterministic way, granting technological change the capacity to shape and reshape workplaces and the jobs and skills of workers.[24] Indeed, the literature around technological change and the future of work is replete with grandiose claims about the imminent reorganisation of entire industries, and even national economies.[25] These claims are often made with limited empirical backing, however. Indeed, an enduring lesson from industrial sociology examining technological change and industrial development is that the managerial literature frequently underestimates the technical and social impediments to the integration of new technologies in workplaces and industry.[26] In contrast, this chapter holds technological change in the workplace – and industrial development more broadly – to be the result of human decision-making mediated by the institutions, politics and economics of the society in which they are deployed, and subject to social struggles between various social actors at multiple levels.[27]

Managerial approaches to these struggles are shaped by tensions between the development of productive forces, on the one hand, and the need to maintain control of the workplace on the other.[28] Driven by the need to compete in competitive markets, capitalist firms are forced to continually develop their productive capabilities through investment in new technologies that help drive productivity improvements.[29] At the same time, in order to maintain control of the means of production, firms must hold a tight rein over their workforces and the work processes they perform. This generates a tension between empowering workers through training and upskilling for greater efficiency on the one hand, and maximising managerial control of the workplace through deskilling and Taylorisation of work processes on the other.[30] At the level of industrial development, this tension presents a risk to the capacity of firms to integrate new technologies. Despite coherent union arguments around industry and workforce development through upskilling and worker empowerment, firms are incentivised to maximise workplace control for short-term profit generation, rather than expanding the capabilities of the workforce through upskilling and worker empowerment.

Examining Australia's two industrial development priorities using this conceptual framework provides us with the analytical tools to explain some of the shortcomings of the naval shipbuilding program and potential risks to the net-zero transformation. In short, the managerial drive for short-term profit maximisation through the intensification of workplace control is jeopardising the broader development of Australia's productive capabilities. This manifestation of capitalist inefficiency is restricting the development of Australian industry both in the defence sector and regarding opportunities to develop new green industrial sectors, placing the government's industrial policy ambitions at risk.

Union approaches: Industrial development through worker empowerment

Here we discuss the recent experiences of trade unions – the AMWU and ASFU – in advocating and campaigning for industrial and workforce development to accompany major acquisitions of defence equipment, and the lessons that can be drawn from this advocacy for the new green industry policy. Specifically, we examine efforts to build a coordinated workforce development programme around Australia's naval shipbuilding programme. The AMWU and ASFU have pursued these issues as a major campaigning focus for several reasons. First, sovereign capability in defence manufacturing can help pursue an independent and peace-oriented foreign policy. Second, major defence manufacturing projects have historically been a source of well-paid, secure, unionised jobs. Third, major government projects such as naval shipbuilding can have strong, positive "spillover" effects, helping develop workforce and industrial capabilities for other manufacturing sectors, such as renewable energy infrastructure.

The AMWU has a long history of working to connect industry policy strategies with upskilling demands at the point of production.[31] This has involved simultaneously advocating policies that promote industry and workforce development at a macro level, and workplace outcomes that upskill and empower workers on the shopfloor. These have included highly developed union strategies surrounding the integration of new technologies across industries and in specific workplaces. Unions argue that these two areas – industry and workforce development, and worker upskilling and empowerment at the point of production – operate as a coherent whole which would facilitate the efficient development of Australia's productive capabilities. Below, we argue that this agenda has been significantly stifled by short-term profit maximisation and a managerial drive to control the workforce.

Today the AMWU and ASFU consistently advocate for workforce development as an essential component of industry policy both to enable defence manufacturing and build a wider manufacturing ecosystem. In its submission to the Defence Strategic Review, the AMWU was critical of the Department of Defence and private defence contractors' workforce development failures, particularly a lack of resources provided to recruitment, investment in skills and training, and succession planning to secure a pipeline of skilled workers to replace an ageing workforce.[32] The AMWU had identified these concerns as early as 2014, when it wrote that failure to adequately plan for workforce development would make major workforce capability issues and skills shortages inevitable.[33] The unions have argued that the Department of Defence and private employers have a narrow understanding of the capabilities needed to sustain industrial capacity for the defence sector, which lead to inadequate fixes being proposed.[34] Among the issues identified by the ASFU were inadequate, often incoherent workforce planning; poor working conditions that lead to high labour turnover; a failure to invest adequately in apprenticeships, traineeships and graduate positions; and generalised hostility towards union advocacy for the upskilling and reclassification of workers. We now unpack some of the specific policy and workplace solutions advocated by the unions, and managerial objections to these proposals. Many of these proposals are centred around the naval shipbuilding precinct in South Australia but are representative of broader union positions around workforce development and industry coordination.

Tripartite workforce mapping and planning

Unions advocate for a tripartite (government, industry and union) workforce planning process.[35] This workforce development programme would involve the mapping of workforce requirements spanning from macro trends (for example, mechanical fitters and machinists and electricians are consistently among the top ten "labour shortages" identified by Jobs and Skills Australia[36]) through to micro skill trends within workplaces. This workforce mapping would involve establishing a series of occupational profiles for the varying job roles in Australia's shipbuilding yards (e.g., boilermakers, electricians, engineers etc.) to identify the skills, tasks and industrial job classifications required for current and future manufacturing projects. Unions propose this would involve deep, cooperative efforts to map the workforce requirements of Australia's future shipbuilding projects and how future needs can be met through training on the ongoing naval shipbuilding projects.[37] Mapping through occupational profiling would facilitate workforce planning and the development of a pipeline of workers trained to the highest possible standards, meeting the competency requirements of future shipbuilding projects.

While employers have not rejected the need for workforce and skills mapping at a general level, many have since pushed back against the idea of working with unions to conduct this mapping – especially at the workplace level. Specifically, employers have consistently pushed back against union efforts to map skills utilisation and training requirements at shopfloor level. Work classification structures, contained in the Manufacturing and Associated Industries and Occupations Award and enshrined in many shipbuilding enterprise agreements, provide a framework for this mapping and the industrial recognition of workers' skills. The framework is designed to incentivise ongoing upskilling for both workers and employers and build an adaptable and dynamic workforce. Under the framework, skills required to perform production tasks are systematically mapped and linked to an industrial mechanism for enforcement. In this way, they provide workers and their unions a lever to influence how production is organised and push for more complex, autonomous work processes. Managers appear to view these enforceable structures as a direct challenge to the frontier of managerial control. Employers have pushed back against enshrining these classification structures into enterprise agreements during bargaining, and union organisers and delegates are frequently obstructed by managers when attempting to conduct occupation profiling and classification activities on the shopfloor in naval shipbuilding workplaces.

Continuous shipbuilding tied to workforce development obligations

The Australian naval shipbuilding programme has long been haunted by what has been dubbed a potential "valley of death."[38] This term refers to the loss of skilled workers and know-how caused by a downturn or gap in production, as a lack of employment options forces skilled workers to seek employment in competing sectors and no investment in training new workers occurs. Such a valley of death can be so costly that modelling from other jurisdictions has found that building ships that would never be used, just to maintain the skilled workforce, would cost less than allowing the workforce to disperse and then reassemble or rebuild it.[39] For this reason, unions, employers and government agree that continuous naval shipbuilding is necessary to make the programme sustainable.

Within this programme of continuous shipbuilding, the ASFU and AMWU argue that government issued defence contracts should come with strong conditionality around their treatment of workers and commitment to local supply chain development. This includes requirements around workforce training to address the skills shortages frequently identified by employers. Unions have consistently advocated for employers to commit to ambitious training requirements when working on government projects. Most tangibly, this includes mandatory ratios for tradespeople and apprentices on any project receiving government funding. In recent submissions, the ASFU and AMWU have advocated for a modest ratio of one apprentice to every ten tradespeople.[40] Additionally, unions have pushed for quotas of apprenticeships and training opportunities for Indigenous Australians and women. While such apprenticeship ratios and quotas have been included in some defence sector contracts, these requirements are not always being met, and a lack of meaningful enforcement mechanisms limits the capacity of unions and governments to force employers to comply.

Vocational training within recognised frameworks

Unions have regularly advocated for on-the-job training to be complemented through Technical and Further Education (TAFE) institutions (Australia's government-funded vocational education system of Registered Training Organisations [RTOs]) based learning. TAFE can provide experience-based and classroom-based training that reflects cutting-edge manufacturing practices, including working with new and emerging technologies. In the context of the South Australian shipbuilding precincts, unions have campaigned for a TAFE campus to be established in or near the precinct – the Engineering Centre of Excellence – to train new workers and upskill the existing workforce to facilitate the integration of new technologies.[41] This Centre would be a training resource for all producers in the shipbuilding and defence supply chains, not only the prime contractors. Importantly, training packages provided through the Centre would be consistent with the Manufacturing and Engineering Training Package in the Australian Qualifications Framework (AQF). This would mean that skill sets taught at the Centre are recognised across the manufacturing industry, ensuring that qualifications are transferable between employers and facilitating positive spillovers of skilled workers into other sectors. Additionally, the skills taught at the Centre would be recognised in the Manufacturing and Associated Industries and Occupations Award, ensuring workers have an industrial mechanism for enforcing the recognition of their skills.[42]

The idea of a centralised training facility of this kind has been met with mixed responses by employers. On one level they appear receptive to the idea of a joint resource for the development of the skilled workforce they require to meet their production objectives. However, the broad-based, industrially recognised training programmes advocated by unions also present risks to employers' control interests. Here, shipbuilding employers have sometimes favoured micro-credentialing offered by private RTOs.[43] These micro-qualifications are often only recognised by a single business, therefore tying workers to their employer. Furthermore, they are not recognised in the AQF or the relevant Award, making it more difficult for unions to enforce recognition of worker upskilling in the industry.[44]

Precinct industrial agreements

Finally, unions have advocated for precinct-wide industrial agreements to avoid a "race to the bottom" whereby employers are incentivised to deskill workers to undercut competitors.[45] The major defence hardware acquisitions flagged in the DSR (missiles and submarines) represent the most complex public procurement projects in Australia's history. Unions propose maximising positive spillovers from these projects by investing in transferable skills and qualifications frameworks mapped to Australian standards and guaranteeing their portability between firms within relevant precincts to ensure consistency of training across employers. Unions have argued that this requires precinct-wide multi-employer industrial agreements at major defence industry precincts to which all firms and employers are party. A consistent classification structure, mapping skills against tasks and corresponding rates of pay in defence manufacturing, would facilitate the movement of workers between employers while taking wage competition out of the equation. This would also address a frequent complaint of employers that their tradespeople are "poached" upon completing their apprenticeship by other employers who do not invest in skills and training. Importantly, the precinct agreements must facilitate trade union involvement in workforce development through the capacity to conduct reclassification inspections for workers when appropriate.

Perhaps unsurprisingly, defence employers – like the business community more broadly – have consistently advocated against multi-employer bargaining of this kind.[46] Innes Willox, Chief Executive Officer of the Australian Industry Group, of which the major shipbuilding and defence contractors are members, said that multi-employer bargaining would "hand unions an unreasonable level of control over the content of agreements."[47] While multi-employer bargaining in the form of precinct agreements appears to point a way beyond the "race to the bottom" on skills and training in the defence manufacturing sector, managerial control concerns remain a significant hurdle to realising a more efficient and equitable model of production.

Capitalist inefficiency, workforce development and lessons for green industrial policy

If Australian industry is to integrate complex technologies like nuclear-powered submarines or "green metal" production, significant workforce development will need to occur, but employers' control concerns currently represent a significant handbrake on progress in this area. As discussed earlier, employers in Australia's defence manufacturing sector share the unions' high-level concerns about workforce development and developing a wider manufacturing ecosystem around defence industrial precincts. Indeed, representatives of both primes and small- and medium-sized enterprises in defence manufacturing supply chains have confided in the AMWU that a lack of skilled workers is their primary business risk. However, while some aspects of the ASFU's plan for workforce and industrial development have been embraced by employers, others have been rejected or obstructed.

In a notable example, a major defence contractor has engaged at high level with the ASFU around the development of a workforce mapping plan based on occupational profiles. This has involved the kind of deep engagement between labour and capital required for such workforce planning to succeed. However, the same firm has strongly opposed the implementation of a competency-based classification structure based on agreed occupational profiles in the process of enterprise bargaining. This is despite senior managers conceding that such a system would facilitate and encourage the much-needed upskilling of workers. At another level, however, human resource executives and some production managers have expressed concern that such a system is a grab for power by the unions, as the classification structures would give workers more control over their labour process. This compels them to obstruct paid ASFU officials from access to worker inductions, which minimises worker interaction with avenues for potential union representation, despite such access being agreed upon at executive level to fulfil strategic workforce development goals held in common by labour and capital.

What can explain this apparent contradiction? In this instance, senior managers are primarily concerned with the development of workforce capabilities and increasing efficiency through the upskilling of workers. Their high-level strategic outlook views the potential shortage of skilled workers as a major risk, strengthening their willingness to engage with workers and unions to develop an upskilling agenda that meets the needs of their projects. On the shopfloor, however, managers are primarily concerned with maintaining control of the workplace and are loath to empower unionised workers with more skills and autonomy. Worried that a competency-based classification structure will upskill workers in ways that grant them high levels of control over the workplace, they have opposed the move. In other words, the senior managers are concerned with the development of the productive forces, while other managers – especially human resource managers – are primarily motivated by maintaining control of the workplace. The intra-organisational bargaining between different factions of management has essentially resulted in a stalemate where little has been achieved.

These contradictions lead us to the notion of capitalist inefficiency. Union ambitions for worker upskilling and broader industrial development have been stifled by the tensions placed on management by the need to develop their productive capabilities on the one hand and control their workforce on the other. This has immediate implications for the Australian government's defence industry ambitions. Most notably, it appears increasingly likely that the largest defence investment in the country's history – the acquisition of nuclear submarines – will see Australia send billions of dollars to the USA and UK, due in part to the lack of industrial capability to build them domestically. This tension has also stifled positive spillovers into other sectors of the manufacturing industry.

The shortcomings of Australia's naval shipbuilding industrial policy point to significant risks for the federal government's "Net Zero Transformation Stream" of FMiA. The government hopes it will drive Australia's energy transition while generating tens of thousands of new well-paid manufacturing jobs. Importantly, the government hopes green industrial policy can help it end the "climate wars" by reframing Australia's energy transition as one that will generate new economic opportunities for the country's legacy fossil fuel regions. To realise these goals will require a coordinated and place-based workforce development programme of the kind Australia's naval shipbuilding programme has struggled to achieve.

Unions, including those involved in the campaigning and advocacy around shipbuilding workforce development described above, have begun arguing for a similar policy agenda around the green industrial transition. For example, they have demanded strong labour conditionality around government investment, including investment in skills and training, as part of a broader suite of policies designed to maximise local industry and workforce development. Concerningly, in its initial phases, governmental and employer approaches to union demands for coordinated workforce development appear to be following a similar pattern to defence. The government has resisted union efforts to enshrine strong labour conditionality mechanisms in FMiA legislation. At the same time, employers in existing green industry projects have resisted union efforts to bargain for upskilling and training while simultaneously bemoaning a lack of skilled workers. Whether or not unions and government can pressure employers to look beyond immediate control concerns to maximise the development of green industry in Australia remains to be seen. However, the country's ability to navigate the polycrisis through the deployment of industrial policy will partly hinge on the answer to this question.

Notes

1 Tooze, *Defining Polycrisis*.
2 Dean, Rainnie, Stanford, and Nahum, "Industrial Policy-making after COVID-19."
3 Department of Finance, *A Future Made in Australia Fact Sheet*.
4 Australian Treasury, *Future Made in Australia National Interest Framework*.
5 Tooze, *Defining Polycrisis*.
6 Ide, "Climate Change and Australia's National Security," 26.
7 Charlton, *Dragon's Tail*.
8 Fernandes, *Sub-imperial Power*.
9 Fernandes, *Sub-imperial Power*, 29.

10 Department of Defence, *Defence Strategic Review.*
11 ASPI, *The Cost of Defence 2020–2021 Part 2*, 68–69.
12 Greene, "Defence Reveals 50 per cent 'Contingency' for Cost Overruns."
13 Department of Defence, *Defence Strategic Review 2023.*
14 Department of Defence, *Defence Strategic Review 2023.*
15 Greene, "First Newly Built Nuclear-powered Submarine under AUKUS Likely to be Sold in 2038."
16 Hurst, "Australia's $4.6bn Aukus Funding."
17 Karp, "Plan to Build Aukus Submarines in Adelaide is 'a Fairytale' and 'Pork Barrelling'."
18 Burke, "On the Way Out."
19 See also Chapter 7 in this volume, Carr and Larkin, "Energy Transitions in a Metallurgical Coal Region."
20 Crowly, "Climate Wars, Carbon Taxes and Toppled Leaders."
21 Cassie, *Collie Just Transition Plan*; Hunter Jobs Alliance, *About the Hunter Jobs Alliance.*
22 Graham and Havas, "Comparing and Ranking the Global Cost of Green Industrial Electricity."
23 Department of Finance, *A Future Made in Australia Fact Sheet.*
24 See, for example: Gregolinska et al., "Capturing the True Value of Industry 4.0."
25 See, for example: Susskind, *A World without Work.*
26 Will-Zocholl, "Virtual Temptations."
27 Hall, "Renewing and Revising."
28 Vidal, "Contradictions of the Labour Process."
29 Hall, "Renewing and Revising."
30 Vidal, "Contradictions of the Labour Process."
31 Frenkel, "The Australian Metalworkers' Union and Industrial Change."
32 AMWU, "Submission to the Defence Strategic Review."
33 AMWU, "Submission to the Joint Standing Committee on Foreign Affairs, Defence and Trade."
34 ASFU, "Closing Australia's Submarine Capability Gap."
35 ASFU, "Building Australia's Future."
36 Jobs and Skills Australia, "Skills Priority List Key Findings Report 2023."
37 Department of Defence, "Naval Shipbuilding Plan."
38 Brookes, "Australian Shipbuilders Facing 'Valley of Death' Due to Nuclear Subs Uncertainty."
39 Davies and Mugg, "Continuous Naval Shipbuilding in the UK."
40 ASFU, "Building Australia's Future."
41 ASFU, "Building Australia's Future."
42 Manufacturing and Engineering Industry Reference Committee, "Manufacturing and Engineering Training Package (MEM) Essential Features Guide."
43 AiGroup, "2022 Skills Survey."
44 Nicholson, "There Is Such Thing as Unskilled Work."
45 ASFU, *Building Australia's Future.*
46 Forsyth and McCrystal, "The Potential Impact of the Fair Work Amendment (Secure Jobs, Better Pay) Act 2022 on Collective Bargaining in Australia."
47 AiGroup, "Welcome Amendments but IR Bill Remains Fatally Flawed."

Bibliography

AiGroup. *Welcome Amendments but IR Bill Remains Fatally Flawed.* Australian Industry Group, 7 November 2022. www.aigroup.com.au/news/media-centre/2022/welcome-amendments-but-ir-bill-remains-fatally-flawed/

AiGroup (Australian Industry Group). *2022 Skills Survey: Listening to Australian Businesses on Skills and Workforce Needs*. Australian Industry Group, 2022. www.aigroup.com.au/news/reports/2022/2022-skills-survey-listening-to-australian-businesses-on-skills-and-workforce-needs

AMWU. *Submission to the Defence Strategic Review*. Melbourne: Australian Manufacturing Workers' Union, 2022.

AMWU (Australian Manufacturing Workers' Union). *Submission to the Joint Standing Committee on Foreign Affairs, Defence and Trade into Government Support for Australian Defence Industry Exports*. Melbourne: Australian Manufacturing Workers' Union, 2014.

ASFU. *Building Australia's Future: The Enduring Workforce and Industry Development Benefit of an Australian Marine Engineering Build*. Melbourne: Australian Shipbuilding Federation of Unions, 2023.

ASFU (Australian Shipbuilding Federation of Unions). *Closing Australia's Submarine Capability Gap*. Melbourne: Australian Shipbuilding Federation of Unions, 2022.

ASPI (Australian Strategic Policy Institute). *The Cost of Defence 2020–2021 Part 2: ASPI 2020–21 Defence Budget Brief*. Canberra: Australian Strategic Policy Institute, 2021.

Australian Treasury. "Future Made in Australia National Interest Framework." *Australian Government Treasury Advice*, May 2024. https://treasury.gov.au/publication/p2024-526942

Brookes, Joseph. "Australian Shipbuilders Facing 'Valley of Death' Due to Nuclear Subs Uncertainty." *AuManufacturing*, 1 March 2022. www.aumanufacturing.com.au/australian-shipbuilders-facing-valley-of-death-due-to-nuclear-subs-uncertainty

Burke, Paul J. "On the Way Out: Government Revenues from Fossil Fuels in Australia." *Australian Journal of Agricultural and Resource Economics* 67 (2023): 1–17.

Cassie, Alex. "Collie Just Transition Plan." *Search Foundation*, May 2021. www.search.org.au/collie_transition_plan

Charlton, Andrew. *Dragon's Tail: The Lucky Country after the China Boom*. Melbourne: Black Inc., 2014.

Crowly, Kate. "Climate Wars, Carbon Taxes and Toppled Leaders: the 30-year History of Australia's Climate Response, in Brief." *The Conversation*, 15 October 2021. https://theconversation.com/climate-wars-carbon-taxes-and-toppled-leaders-the-30-year-history-of-australias-climate-response-in-brief-169545

Davies, Andrew and James Mugg. "Continuous Naval Shipbuilding in the UK." *Australian Strategic Policy Institute*, May 2016. www.aspistrategist.org.au/continuous-naval-shipbuilding-in-the-uk/

Dean, Mark, Al Rainnie, Jim Stanford and Dan Nahum. "Industrial Policy-making after COVID-19: Manufacturing, Innovation and Sustainability." *The Economic and Labour Relations Review* 32, no. 2 (2021): 283–303.

Department of Defence. *Naval Shipbuilding Plan*. Canberra: Department of Defence, Australian Government, 2017. Accessed March 2024: www.defence.gov.au/business-industry/naval-shipbuilding/plan

Department of Defence. *Defence Strategic Review*. Canberra: Department of Defence, Australian Government, 2023.

Department of Finance. "A Future Made in Australia Fact Sheet." *Budget 2024 Fact Sheet*. Canberra: Department of Finance, Australian Government, 2024. https://budget.gov.au/content/factsheets/download/factsheet-fmia.pdf

Fernandes, Clinton. *Sub-imperial Power: Australia in the International Arena*, Melbourne: Melbourne University Press: 2022.

Forsyth, Anthony and Shae McCrystal. "The Potential Impact of the Fair Work Amendment (Secure Jobs, Better Pay) Act 2022 on Collective Bargaining in Australia: Reviewing the new Multi-employer Bargaining Provisions and other Measures to Promote Bargaining." *Journal of Industrial Relations* 65, no. 4 (2023): 386–402.

Frenkel, Stephen. "The Australian Metalworkers' Union and Industrial Change: A Labour Movement Offensive." In *New Technology and Industrial Relations*, edited by R. Hyman and W. Streeck, 233–46. Oxford: Basil Blackwell, 1988.

Graham, P.W. and Havas, L. "Comparing and Ranking the Global Cost of Green Industrial Electricity." In *25th International Congress on Modelling and Simulation*, Darwin, NT, 9–14 July 2023.

Greene, Andrew. "Defence Reveals 50 per cent 'Contingency' for Cost Overruns inside $368 Billion AUKUS Project." *ABC News*, 28 April 2023a. www.abc.net.au/news/2023-04-28/defence-reveals-50-per-cent-contingency-for-cost-overruns-inside/102275806

Greene, Andrew. "First Newly Built Nuclear-powered Submarine under AUKUS likely to be Sold in 2038, US Admiral Reveals." *ABC News*, 9 November 2023b. www.abc.net.au/news/2023-11-09/aukus-submarine-sales-timelines-revealed/103083780

Gregolinska, Ewelina, Rehana Khanam, Frédéric Lefort and Prashanth Parthasarathy. *Capturing the True Value of Industry 4.0.* Reports: McKinsey and Company, 2022. www.mckinsey.com/capabilities/operations/our-insights/capturing-the-true-value-of-industry-four-point-zero

Hall, Richard. "Renewing and Revising the Engagement between Labour Process Theory and Technology." In *Working life: renewing labour process analysis. Thompson*, edited by Paul Thompson and Chris Smith, 159–82. Basingstoke: Palgrave Macmillan, 2010.

Hunter Jobs Alliance. "About the Hunter Jobs Alliance." https://hunterjobsalliance.org.au/about/

Hurst, Daniel. "Australia's $4.6bn Aukus Funding to Help Create More than 1,000 Jobs in UK, Rolls-Royce Says." *The Guardian*, 25 March 2024. www.theguardian.com/world/2024/mar/25/australia-aukus-funding-uk-jobs-rolls-royce

Ide, Tobias. "Climate Change and Australia's National Security." *Australian Journal of International Affairs* 77, no. 1 (2023): 26–44.

Jobs and Skills Australia. *Skills Priority List Key Findings Report 2023*. Canberra: Jobs and Skills Australia, 2023. www.jobsandskills.gov.au/publications/skills-priority-list-key-findings-report-2023

Karp, Paul. "Plan to Build Aukus Submarines in Adelaide is 'a Fairytale' and 'Pork Barrelling', Alexander Downer Says." *The Guardian*, 30 October 2023. www.theguardian.com/australia-news/2023/oct/30/aukus-submarine-deal-built-adelaide-alexander-downer

Manufacturing and Engineering Training Package (MEM). "Essential Features Guide." March 2021.

Nicholson, Daniel. "There Is Such Thing as Unskilled Work." *Australian Fabians*, 25 September 2023. www.fabians.org.au/there_is_such_thing_as_unskilled_work

Susskind, Daniel. *A World Without Work. Technology, Automation, and How We Should Respond.* London: Allen Lane, 2020.

Tooze, Adam. *Defining Polycrisis – From Crisis Pictures to The Crisis Matrix*. Chartbook #130: June 2022. https://adamtooze.com/2022/06/24/chartbook-130-defining-polycrisis-from-crisis-pictures-to-the-crisis-matrix/

Vidal, Matt. "Contradictions of the Labour Process, Worker Empowerment, and Capitalist Inefficiency." *Historical Materialism* 28, no. 2 (2019): 170–204.

Will-Zocholl, Mascha. "Virtual Temptations: Reorganising Work under Conditions of Digitisation, Virtualisation and Informationisation." In *The New Digital Workplace: How Technologies Revolutionise Work*, edited by Birgit Briken et al., 62–88. Basingstoke: Palgrave Macmillan, 2017.

11

FROM OCCUPATIONAL STEERING TO OCCUPATIONAL CITIZENSHIP?

Understanding Career Pathways after the Demise of Australian Automotive Manufacturing

Tom Barnes

Introduction

2024 marks ten years since Australia's last passenger car manufacturers made public their decision to close local operations. In February 2014, General Motors Holden and Toyota announced plans to wind down manufacturing facilities in Adelaide and Melbourne over three years, following Ford's announcement during the previous year to close production in Melbourne and the nearby city of Geelong. The last Australian-manufactured car – the Commodore – rolled off Holden's Adelaide assembly line in October 2017. A decade down the road, these decisions remain a source of debate and angst about the long history, socioeconomic role and uncertain future of manufacturing in Australia. This pattern reflects the historical trajectories of deindustrialisation experienced in other high-income countries in Europe and North America, including mass job losses and destructive impacts on working-class communities.[1] For auto manufacturing, a particular challenge for Australia – in contrast to the international experience – was the disappearance of the industry *in its entirety*.

Despite this disappearance of auto manufacturing jobs, one positive outcome was the comprehensive nature of the assistance for workers, supply chain businesses and communities affected by the closures. These measures meant that only a minority of supply-chain manufacturers closed. Despite initial projections of direct and indirect job losses between 40,000 and 200,000,[2] government data suggests that around 14,000 jobs had been lost 12 months after the industry had closed.[3] Redeployment measures into non-manufacturing roles such as sales and distribution, testing, and aftermarket products, as well as Federal and State Government subsidies for suppliers, meant that total job losses were much lower than initially feared.

DOI: 10.4324/9781003512240-14

This chapter focuses on the practice of labour market assistance rolled out as part of these remedial measures, a practice that widely discouraged workers from pursuing manufacturing careers after the closure of car manufacturing. It is therefore highly surprising, the chapter shows, that thousands chose to effectively ignore this formal advice by *maintaining careers in manufacturing occupations*. To explain this unexpected outcome, this chapter deploys the concept of occupational citizenship, defined as the right to practice one's occupation, one's calling in life, or to belong to a community of shared practice.[4] Occupational citizenship, the chapter argues, provides integrity and meaning to individuals' workplace capabilities and, ultimately, also to their sense of subjective wellbeing. This concept intersects with both the personal meaningfulness of workers' careers and the geography of occupational demand, to explain why so many ignored advice from labour market intermediaries, government programmes and industry experts.

Findings from this chapter point to important lessons for policymakers in the design of labour market programmes for a range of industries in manufacturing and resource-intensive sectors that have grappled with job losses due to economic, technological and environmental change. The chapter argues that the concept of occupational citizenship provides insights for "just transitions" which can look to nurture, rather than jettison, the wealth of occupational skills and experiences concentrated in these sectors. It proceeds in the following sections: in the second section below, the nature of this problem is placed in the context of labour market outcomes for retrenched auto workers. The third section outlines industry and occupational outcomes for workers, before the fourth section outlines the contrast with formal programmes of labour market assistance which often explicitly encouraged workers to abandon extant occupations through a process of "occupational steering." The fifth section of the chapter suggests that the concept of occupational citizenship can address the gap between labour market planning and labour market outcomes for auto workers, before concluding the chapter.

Job market outcomes for auto workers after closure

Evidence from extant studies suggests that labour market outcomes were relatively poor following the closure of the car manufacturing industry.[5] By 2016, shortly prior to most of the closures, over 90 percent of auto workers were in permanent fulltime jobs. 12 months after the final closures, large numbers of workers were unemployed or employed in insecure jobs.[6] A government-sponsored national study found that 15 percent of workers were unemployed jobseekers 12 months after closure, 70 percent were in paid work, 6 percent were retired, and 9 percent had withdrawn temporarily from the labour force.[7]

A further longitudinal study of 1277 auto workers which commenced in mid-2020 – almost three years after closure – found that 71 percent were in paid work, 6 percent retired, 11 percent had withdrawn temporarily from the labour force, and 12 percent were unemployed.[8] While these outcomes look poor, Australia was in a period of lockdowns and deep recession in the context of the Covid-19 pandemic. During this period, like many similar countries around the world, Australia pursued a policy of community lockdowns which shut down labour markets. By mid-2022, approximately five years post-closure, 75 percent of workers were in paid work, 9 percent retired, 12 percent withdrawn from the labour force, and 4 percent unemployed (n = 783).[9]

Despite these poor labour market outcomes, many workers retained a strong desire to retain their occupational profile and status. In one early study, 49 percent said it was "important" or "essential" that they stayed "in the same or a similar occupation." Nearly two-thirds (65 percent) were still looking for jobs in the manufacturing sector in the 12 months after the closures.[10] In industry terms, 35 percent of workers remained in the manufacturing industry 12 months after the closure of the auto manufacturing industry. This was easily the largest industry share of jobs post-closure, followed by 13 percent in transport, postal and warehousing jobs, and 9 percent in construction, as well as some employment in healthcare and social assistance. Nevertheless, these data prompted the conclusion that "manufacturing [was] still a significant employer" for retrenched auto workers.[11]

Three years after closure, the proportion of workers in manufacturing was *even higher*. According to data from the *Future Work Future Communities* (FWFC) research team (Wave 1, 2020, n = 824),[12] nearly half of workers in new jobs (46 percent) were in the manufacturing industry. The next largest share was transport, postal and warehousing employment (8 percent), followed by construction (7 percent), healthcare and social assistance, and public administration and defence (5 percent each).

In occupational terms, the same dataset shows that, at the time of the closures, 39 percent of workers in the auto industry were manual process workers, operators or labourers, 28 percent manual craft workers, 10 percent clerical or service workers, 15 percent professionals and 9 percent managers. The distinction between manual process work and manual craft work typifies shopfloor employment in auto manufacturing; the former refers to a range of less formally skilled and semi-skilled occupations that usually require little formal education – for example, a secondary school certificate and some on-the-job training – encompassing a range of manual process jobs such as product assembly, quality control or materials handling. The latter is comprised of skilled and semi-skilled trades and technical occupations that usually require at least a Certificate III in formal vocational training, or the equivalent of three years' on-the-job training, and often an associate degree or diploma; key examples include spray painters, electricians, welders, fitters and turners, toolmakers and engineering technicians.[13]

In terms of new jobs three years after closure, if we control for those workers who were unemployed, withdrawn from the labour force, and a small minority of self-employed workers, we find that 34 percent of workers were still manual process workers, 19 percent were manual craft workers, 20 percent were service or clerical workers, 20 percent professionals and 7 percent managers. In fact, workers' occupational groups pre- and post-closure are significantly correlated (Spearman's $\rho = 0.387$, $p < 0.001$, $n = 1277$). If we exclude those outside the labour force and the minority of self-employed workers, these groups are significantly *and strongly* correlated (Spearman's $\rho = 0.658$, $p < 0.001$, $n = 824$). If we re-run the correlation as a Chi-square test, they remain significantly associated ($X(16) = 930.830$, $p < 0.001$).[14] Whichever statistical test we apply, results suggest there has been strong continuity in intra-occupational pathways despite poor labour market outcomes overall. Thus, in both industry and occupational terms, findings suggest that the largest share of workers remained as manual workers in manufacturing-based jobs.

Contrasting labour market outcomes with labour market programmes

Over half of workers surveyed undertook training or re-education prior to job loss. Under various programmes of labour market assistance before the closure, all workers were entitled to four stages of labour market assistance.[15] First, workers received basic information about the closures and access to careers counselling as well as financial planning and other services. Second, workers received assistance to develop individual "Career Transition Plans" (CTPs). A CTP represented a written agreement between a worker and a labour market agency which outlined the type of job the worker would target, the skills and qualifications needed to apply, and gaps in the worker's resume to be addressed. Third, workers with gaps in skills and qualifications sanctioned under a CTP could access funding to subsidise courses with vocational training providers and educational institutions. Finally, workers were entitled to assistance with job matching or job placement in the lead-up to closure. In addition to these measures, most workers had access to significant redundancy payments after many years of service.

Alongside programme delivery in the workplace, the Federal Government directed retrenched workers to the Jobactive system, which was established in the 1990s and, like welfare-to-work or "workfare" systems internationally, designed to move recipients off unemployment benefits into paid work as quickly as possible. Even though most auto workers did not qualify for unemployment benefits due to their redundancy payouts or other factors such as spousal income, they were eligible to receive labour market assistance through the network of private agencies, or Jobactive Providers (JPs), within this system. Under a specially arranged "Structural Adjustment Program," JPs were incentivised to find training or jobs for auto workers in return for placement fees.[16] At the time of closures, around 2,700 workers had registered for assistance with JPs.[17]

The Federal Government also grouped auto workers under a service delivery banner known as *What's Next?* For job searching, *What's Next* advised workers to:

> Be prepared to step out of your comfort zone and do work that is different to anything you've done before … Understand the different steps involved in searching for a job and develop a plan or strategy around how to best tackle these. Remember, your plan is not set in concrete – you can update or change it at any time.[18]

This approach thus encouraged workers to develop CTPs by adopting entrepreneurial outlooks and behaviours. In the formulation of a CTP prior to closure, workers were encouraged to choose from a wide range of in-demand occupations. Once preferences were narrowed down, workers were assisted to make a case for training subsidies via a formal CTP. This usually involved selecting a single job or occupational goal and assessing the formal training requirements, qualifications or credentials needed to apply. In general, workers had to choose a single pathway. "Your choice [must be] only one… [Once] you choose, that's it," as explained by one worker.[19]

In interviews with auto workers and case managers, the process of selection and codification was described repeatedly as "steering" or "guiding."[20] Such redirection encouraged workers to abandon long-term occupations, to lower job quality expectations, and to reorient to high-demand vocations, as highlighted by one case manager: "The important thing is managing expectations. This is … the primary focus."[21] Case managers were appointed in different ways. In the three car-makers, case managers were employed to work in company offices, either as secondments from Human Resource (HR) departments or as representatives from third-party skills providers. In the supply chain, case managers were generally appointed by government-sanctioned Registered Training Organisations (RTOs) or consultancies to visit workplaces occasionally. Elsewhere, case managers could be found in JP offices or State Government-funded skills centres. But, regardless of the institutional context, all case managers were required to codify the narrowing of workers' career and occupational goals by drafting CTPs. This was an individual process, crafted through one-on-one repeat consultations with workers. The individualisation of case management was partly a response to the failure of multi-worker seminars – stigmatised as a "classroom-based approach" by one interviewee – to attract workers with language, literacy and numeracy difficulties, including many from culturally and linguistically diverse backgrounds. But this individualised approach also provided a means of nudging workers into high demand manual occupations outside manufacturing by, for example, discouraging workers from "following their mates" (as phrased by one case manager) into popular training destinations. The key to case management was to establish trust and encourage workers to "open up," thereby laying the basis for occupational redirection. The identification of a training- or credential-based pathway into a new job had to satisfy programme managers; all subsidy payments had to be justified through a "realistic" CTP.

One-on-one case management encouraged workers to think "outside the box" of previous career trajectories; to be realistic about moving into an industry or occupation where they were more likely to receive immediate job offers. Under the *What's Next* rubric, workers were encouraged to follow three steps to a new career: "make a start," "get ready," and "make it happen." "Making a start" included acquiring "an understanding of where the jobs are in the Australian labour market, and where they will be in the future," along with the following advice: "Some jobs may not be available in your local area and you may want to consider relocating":

> Visit the place you want to move to in order to find out more about rent, transport and potential employers. Line up some work experience in an industry you are interested in. Join professional associations and do some networking to find out where the jobs are.[22]

As this quote suggests, policymakers encouraged workers to be as mobile as possible in geographical as well as inter-sectoral terms. Labour market programmes included "good news stories" about workers who had successfully transitioned into new occupations. One worker, Santo, had worked at Holden in Adelaide until 2006, when he took a redundancy package. Santo qualified as a heavy vehicle driver, from where he eventually found a job working for a local council. A video about Santo's story reinforces the narrative of an entrepreneurial mindset; of individuals willing to try "new things," as well as the rationale for engaging with as many labour market intermediaries as possible ("At the end of the day, the more certificates you've got, the easier it is to get a job").[23]

Under *What's Next*, workers were actively advised to change occupations under the theme of "Changing Lanes." After being told to step out of their "comfort zone," workers' enthusiasm for trying new jobs through trial and error was tempered by the need to be realistic about the availability of jobs:

> It is easier to get a job if you already have a job and if you can take that initial step into the workforce, even on a part time or casual basis, you will be much better placed to find the long term job that is right for you.[24]

After telling his peers to be upbeat and optimistic about job-hunting, Santo explained that he had to start out as a casual, after leaving the automotive sector before moving into a permanent local government job after a period on a fixed-term contract. Santo's rationale was, "unless you try certain jobs, you don't know what you're going to like."[25] Again, this narrative combined the entrepreneurial imperative – be positive, think outside the box, take risks – with the need to accept that employment might mean accepting the reality of a job outside manufacturing. Yet, as explained in the following section, this cautionary logic did not fit so easily with the labour market aspirations or experiences of workers.

Interviewed workers were generally appreciative of labour market assistance *in the workplace* prior to closure, with participants commonly praising the openness, affability and professionalism of case workers and employers. However, workers were often highly critical of private training providers and typically cynical towards services provided through JPs. In these interviews, JPs were invariably described as "unhelpful," "inflexible" or even "useless." The work-first incentive structures of the Jobactive system meant that case managers in JPs were less sensitive to the mobility restrictions of workers – for example, the desire to find work close to home due to caring responsibilities, housing costs or other social priorities – and commonly left participants feeling isolated by a relatively impersonal system which is better designed to punish welfare recipients than find meaningful employment.[26] But even in other contexts, the entrepreneurial approach of case managers clashed with the desire of workers to transition into new industries and workplaces without abandoning the occupational ethos of their careers, as demonstrated in the following section.

From occupational steering to occupational citizenship

Given the efforts to nudge workers into high-demand manual vocations in sectors such as warehousing, transport, construction and healthcare, the proportion of workers who remained in the manufacturing industry seems remarkably high. Despite poor labour market outcomes overall, workers have evidenced considerable success in branching from auto manufacturing into new manufacturing activities. How do we explain this outcome?

A key answer advanced in this chapter is that thousands of workers who successfully maintained careers in manufacturing-based occupations exhibited what is known as occupational citizenship. Occupational citizenship is a fundamental social goal which reflects the "right to practice"[27] and the "right to belong to a community."[28] This concept draws upon critical interpretations of the human capability framework[29] that frames occupations as bundles of practices, competencies and "functionings" that combine into an individual's potential to pursue their chosen field; that provide the "associational freedom" to participate in one's calling; and that provides a mechanism for meaning and sense-making and, thereby, a road to subjective wellbeing or personal fulfilment.[30]

In this vein, the right to practice can also be understood as an element of solidaristic experience and a drive for continuity in the face of economic and social disruption, one that is necessary for individual self-esteem and self-respect.[31] This role of occupational citizenship in meaning-making, recognition and self-respect for workers expelled from long-term careers helps to explain why so many ignored advice to abandon manufacturing careers. Occupational citizenship intersects with experiences of the manufacturing labour process as meaningful because they enable many workers to independently practise their knowledge and skills, to exercise discretion and judgement over their work, and to fulfil many of their personal interests in a workplace context.[32]

The possibility of workers maintaining the integrity of their occupational skillsets also intersects with the availability of manufacturing jobs in closure-affected regions. While this was geographically uneven after the car plant closures, ongoing opportunities were available in manufacturing roles beyond this industry. This was especially so in south-eastern Melbourne, the base for most auto supply chain employment in the country, and also the base for significant ongoing production the manufacturing of vehicles and equipment for railways, trams, trucks, off-road accessories, electrical products and construction equipment.[33] This was also true for Melbourne's northern and western suburbs, although diverse manufacturing opportunities were also available in other places.

Occupational citizenship intersects with these structural and locational factors to explain workers' career trajectories post-closure. It also represents a new point of "destination" for thousands of auto workers – at least for the large minority who transitioned into new jobs within manufacturing – who had moved away from previous notions of "industrial citizenship," based on the atrophying entitlements of industrial wage labour. This idea frames the occupation as "a sense of a lifetime … of creative and dignifying work around a self-chosen set of activities."[34] An occupation has a jurisdiction

> in which claimed 'rights' can include a monopoly of practice, a structure of public payments, a right to impose work procedures on practitioners, an acquired capacity of self-discipline and regulation, and a measure of influence over recruitment, training and licensing.[35]

The rights incurred via membership of an occupational community invoke ideas of workplace voice; of a sense of bargaining power which is tied to individual self-expression, self-worth and self-esteem.

Conclusion

Alongside place-based labour market demand for occupational skillsets and the meaningfulness that workers attach to the manufacturing labour process, the idea of occupational citizenship helps to explain why so many workers defied efforts to nudge them into careers and occupational pathways outside manufacturing. The apparent refusal of so many workers to follow these carefully laid pathways is an important outcome, not least because of ongoing political debate and controversy about the demise of Australian car manufacturing and ongoing uncertainty about high value-added, technology-intensive manufacturing in the country. Such debate encompasses concerns about Australia's capacity to drawdown, transform or reequip key industries in the wider context of uncertainty about technological disruption and climate change. The experience of automotive closures documented in this chapter suggests that the country has retained much of the occupational and skills-based capacity needed to successfully transition and to address these challenges.

In general, Australian government responses to the automotive closure announcements were highly effective in softening the blow of the industry's demise by helping supply chain companies to diversify production, thereby, saving thousands of jobs. While thousands of jobs *were* lost, total job losses have been far lower than even conservative predictions a decade ago.[36] On the other hand, labour market outcomes for retrenched workers have been predictably poor. Even three years after the final car ran off the assembly line, around 12 percent of workers remained unemployed. While this figure was undoubtedly elevated in the context of the Covid-19 recession, the rate of unemployment among retrenched auto workers was around double the national rate at this time and has remained well above the national level even as the economy has recovered since 2021. Similarly, almost a third of workers (31 percent) in paid employment had insecure working arrangements as casuals or fixed-term employees – a far cry from the relatively stable, permanent fulltime jobs that characterised auto manufacturing for much of the post-World War Two period.

In this context of poor labour market and job quality outcomes, it is remarkable that so many workers remained in manufacturing jobs after the closures. A year after the final closures, almost a third of workers remained in the manufacturing sector; three years after the closures, this proportion had *increased* to nearly half of workers in paid employment. In occupational terms, over half of workers (54 percent) remained in manual trades and occupations. This reality is reflected in the strong statistical correlation between occupational groups before and after the closures.

But this outcome is even more remarkable when we consider that workers were commonly discouraged from pursuing manufacturing careers in the lead-up to the closures. Formal programmes of labour market assistance were predicated upon the codification of career pathways that led *beyond manufacturing*. Case management of worker transitions was based on a type of occupational redirection or steering. Within the Jobactive system, this orientation was enhanced by the "work first" approach that incentivised job placement agencies to force workers into the first available job opening. Workers were encouraged to accept first-come job offers, regardless of their desirability or occupational "fit."

In the long term, the fact that nearly half of workers still in paid employment remained in manufacturing jobs implies that a significant minority refused to fit the mould cast through these formal mechanisms of labour market assistance. This collective act of subject refusal can be framed, this chapter has suggested, by workers' commitment to their occupational communities; by their adherence to a right to practice. The concept of occupational citizenship captures this phenomenon neatly. At the same time, the hypothetical element of this thesis points to the need for future research which can analyse career and occupational pathways with greater analytical rigour; for research which can quantify relationships between occupational pathways before and after plant closures. Nevertheless, the cogency of occupational citizenship provides a means of framing this research agenda and also provides policymakers with cause to reflect upon the weaknesses, as well as the strengths, of transitional assistance for the automotive industry and its workers. If the notion of a "just transition" implies that workers in industries threatened by closure are to be engaged meaningfully in processes of change, then this must include workers' attachment to their work and to their occupational communities. The capacity of Australia's former auto workers to achieve a degree of success in branching into new manufacturing jobs *despite* the discouragement of state-sanctioned labour market intermediaries is a testament to such attachments. At the same time, it also serves as a timely warning to policymakers to closely integrate the integrity of occupational skillsets into programmes of labour market assistance for industries affected by climate change, new technologies and economic restructuring around the world.

Notes

1 High, *Industrial Sunset*; Hudson, "Rethinking Change in Old Industrial Regions."
2 Productivity Commission, *Australia's Automotive Manufacturing Industry*.
3 Wallis/ACIL Allen, *The Transition of the Australian Car Manufacturing Sector*.
4 Standing, *Work after Globalisation*; Standing, *A Precariat Charter*; Budd, *The Thought of Work*.
5 Barnes, "Pathways to Precarity"; Irving, Beer, Weller, and Barnes, "Plant Closures in Australia's Automotive Industry."
6 Barnes, "Pathways to Precarity."
7 Wallis/ACIL Allen, *The Transition of the Australian Car Manufacturing Sector*.
8 Irving et al., "Plant Closures in Australia's Automotive Industry."
9 Irving et al., "Plant Closures in Australia's Automotive Industry."

10 Barnes, "Pathways to Precarity."
11 Wallis/ACIL Allen, *The Transition of the Australian Car Manufacturing Sector*, 36.
12 Irving et al., "Plant Closures in Australia's Automotive Industry."
13 Barnes, "Pathways to Precarity."
14 The use of Spearman's ρ assumes that occupational groups are ranked ordinally; for example, manual process worker (0), manual craft workers (1), professional (2), and so on. Chi-squared regression assumes that the same groups are similarly categorical but not ranked ordinally.
15 Barnes, "Punish, Protect or Redirect?"
16 Barnes, "Punish, Protect or Redirect?"
17 Wallis/ACIL Allen, *The Transition of the Australian Car Manufacturing Sector*.
18 Department of Jobs and Small Business, *What's Next?*
19 Cited in Barnes, "Punish, Protect or Redirect?" 15.
20 Cited in Barnes, "Punish, Protect or Redirect?" 15.
21 Wallis/ACIL Allen, *The Transition of the Australian Car Manufacturing Sector*, 71.
22 Department of Jobs and Small Business, *What's Next?*
23 Department of Jobs and Small Business, *What's Next?*
24 Department of Jobs and Small Business, *What's Next?*
25 Department of Jobs and Small Business, *What's Next?*
26 Barnes, "Punish, Protect or Redirect?"
27 Standing, *Work After Globalisation*, 262.
28 Standing, *A Precariat Charter*, 5.
29 Nussbaum, *Women and Human Development*; Sen, *Development as Freedom*.
30 Budd, *The Thought of Work*.
31 Honneth, *The Struggle for Recognition*.
32 Laaser and Karlsson 2022, "Towards a Sociology of Meaningful Work."
33 Beer et al., "Place-based Industrial Strategy and Economic Trajectory.".
34 Standing, *Work After Globalisation*, 10.
35 Standing, *Work After Globalisation*, 11.
36 Productivity Commission, *Australia's Automotive Manufacturing Industry*.

Bibliography

Barnes, Tom. "Pathways to Precarity: Work, Financial Insecurity and Wage Dependency among Australia's Retrenched Auto Workers." *Journal of Sociology* 57, no. 2 (2021): 443–63.

Barnes, Tom. "Punish, Protect or Redirect? Synthesising Workfare with 'Spatially Keynesian' Labour Market Policies in Times of Job Loss." *Environment and Planning A* 55, no. 4 (2022): 871–89.

Beer, Andrew, Tom Barnes and Sandy Horne. "Place-based Industrial Strategy and Economic Trajectory: Advancing Agency-based Approaches." *Regional Studies* 57, no. 6 (2023): 984–97.

Budd, John W. *The Thought of Work*. Ithaca, NY: Cornell University Press, 2011.

Department of Jobs and Small Business. *What's Next?* Canberra: Commonwealth Government of Australia, 2018. https://whatsnext.jobs.gov.au/ (accessed 14 September 2018).

High, Steven. *Industrial Sunset: The Making of North America's Rust Belt, 1969–1984*. Toronto: University of Toronto Press, 2003.

Honneth, Axel. *The Struggle for Recognition: The Moral Grammar of Social Conflicts*. Cambridge, MA: The MIT Press, 1995.

Hudson, Ray. "Rethinking Change in Old Industrial Regions: Reflecting on the Experiences of Northeast England." *Environment and Planning A* 37, no. 4 (2005): 581–96.

Irving, Jacob, Andrew Beer, Sally A. Weller and Tom Barnes. "Plant Closures in Australia's Automotive Industry: Continuity and Change." *Regional Studies, Regional Science* 9, no. 1 (2022): 5–22.

Laaser, Knut and Jan C. Karlsson. "Towards a Sociology of Meaningful Work." *Work Employment and Society* 36, no. 5 (2022): 798–815.

Nussbaum, Martha C. *Women and Human Development: The Capabilities Approach.* Cambridge: Cambridge University Press, 2000.

Productivity Commission. *Australia's Automotive Manufacturing Industry*, Inquiry Report 70, 31 March 2014. Canberra.

Sen, Amartya. *Development as Freedom.* Oxford: Oxford University Press, 1999.

Standing, Guy. *Work After Globalisation: Building Occupational Citizenship.* London: Edward Elgar, 2009.

Standing, Guy. *A Precariat Charter: From Denizens to Citizens*, London: Bloomsbury, 2014.

Wallis/ACIL Allen. *The Transition of the Australian Car Manufacturing Sector: Outcomes and Best Practices.* Prepared for Australian Government Department of Education, Skills and Employment. Canberra: ACIL Allen Consulting/Wallis Strategic Market and Social Research, May 2020.

INDEX

For Product Safety Concerns and Information please contact our EU representative GPSR@taylorandfrancis.com
Taylor & Francis Verlag GmbH, Kaufingerstraße 24, 80331 München, Germany

www.ingramcontent.com/pod-product-compliance
Ingram Content Group UK Ltd.
Pitfield, Milton Keynes, MK11 3LW, UK
UKHW020008200926
475702UK00015B/2546
9781032817712